SPEECH TIME-FREQUENCY REPRESENTATIONS

THE KLUWER INTERNATIONAL SERIES
IN ENGINEERING AND COMPUTER SCIENCE

VLSI, COMPUTER ARCHITECTURE AND
DIGITAL SIGNAL PROCESSING

Consulting Editor

Jonathan Allen

Other books in the series:

Logic Minimization Algorithms for VLSI Synthesis. R.K. Brayton, G.D. Hachtel, C.T. McMullen, and A.L. Sangiovanni-Vincentelli. ISBN 0-89838-164-9.
Adaptive Filters: Structures, Algorithms, and Applications. M.L. Honig and D.G. Messerschmitt. ISBN 0-89838-163-0.
Introduction to VLSI Silicon Devices: Physics, Technology and Characterization. B. El-Kareh and R.J. Bombard. ISBN 0-89838-210-6.
Latchup in CMOS Technology: The Problem and Its Cure. R.R. Troutman. ISBN 0-89838-215-7.
Digital CMOS Circuit Design. M. Annaratone. ISBN 0-89838-224-6.
The Bounding Approach to VLSI Circuit Simulation. C.A. Zukowski. ISBN 0-89838-176-2.
Multi-Level Simulation for VLSI Design. D.D. Hill and D.R. Coelho. ISBN 0-89838-184-3.
Relaxation Techniques for the Simulation of VLSI Circuits. J. White and A. Sangiovanni-Vincentelli. ISBN 0-89838-186-X.
VLSI CAD Tools and Applications. W. Fichtner and M. Morf, editors. ISBN 0-89838-193-2.
A VLSI Architecture for Concurrent Data Structures. W.J. Dally. ISBN 0-89838-235-1.
Yield Simulation for Integrated Circuits. D.M.H. Walker. ISBN 0-89838-244-0.
VLSI Specification, Verification and Synthesis. G. Birtwistle and P.A. Subrahmanyam. ISBN 0-89838-246-7.
Fundamentals of Computer-Aided Circuit Simulation. W.J. McCalla. ISBN 0-89838-248-3.
Serial Data Computation. S.G. Smith and P.B. Denyer. ISBN 0-89838-253-X.
Phonological Parsing in Speech Recognition. K.W. Church. ISBN 0-89838-250-5.
Simulated Annealing for VLSI Design. D.F. Wong, H.W. Leong, and C.L. Liu. ISBN 0-89838-256-4.
Polycrystalline Silicon for Integrated Circuit Applications. T. Kamins. ISBN 0-89838-259-9.
FET Modeling for Circuit Simulation. D. Divekar. ISBN 0-89838-264-5.
VLSI Placement and Global Routing Using Simulated Annealing. C. Sechen. ISBN 0-89838-281-5.
Adaptive Filters and Equalisers. B. Mulgrew, C.F.N. Cowan. ISBN 0-89838-285-8.
Computer-Aided Design and VLSI Device Development, Second Edition. K.M. Cham, S-Y. Oh, J.L. Moll, K. Lee, P. Vande Voorde, D. Chin. ISBN: 0-89838-277-7.
Automatic Speech Recognition. K-F. Lee. ISBN 0-89838-296-3.
A Systolic Array Optimizing Compiler. M.S. Lam. ISBN: 0-89838-300-5.
Algorithms and Techniques for VLSI Layout Synthesis. D. Hill, D. Shugard, J. Fishburn, K. Keutzer. ISBN: 0-89838-301-3.

SPEECH TIME-FREQUENCY REPRESENTATIONS

by

Michael D. Riley

AT&T Bell Laboratories

KLUWER ACADEMIC PUBLISHERS
BOSTON/DORDRECHT/LONDON

Distributors for North America:
Kluwer Academic Publishers
101 Philip Drive
Assinippi Park
Norwell, Massachusetts 02061 USA

Distributors for the UK and Ireland:
Kluwer Academic Publishers
Falcon House, Queen Square
Lancaster LA1 1RN, UNITED KINGDOM

Distributors for all other countries:
Kluwer Academic Publishers Group
Distribution Centre
Post Office Box 322
3300 AH Dordrecht, THE NETHERLANDS

Library of Congress Cataloging-in-Publication Data

Riley, Michael D. (Michael Dennis), 1955-
    Speech time-frequency representations / by Michael D. Riley.
        p.   cm. -- (The Kluwer international series in engineering and
computer science ; 63. VLSI, computer architecture, and digital
signal processing)
    Bibliography: p.
    Includes index.
    ISBN 0-89838-298-X
    1. Automatic speech recognition. I.Title.   II. Series: Kluwer
international series in engineering and computer science ; SECS 63.
III. Series: Kluwer international series in engineering and computer
science. VLSI, computer architecture, and digital signal
processing.
TK7882.S65R55 1989
006.4'54--dc19   88-7585
                                                             CIP

Contents

Figures

3   TIME-FREQUENCY FILTERING

4   THE SCHEMATIC SPECTROGRAM

5   A CATALOG OF EXAMPLES

Acknowledgments

This book is based on my Ph.D thesis in E.E.C.S. from M.I.T., submitted in May 1987. I wish to thank Tom Knight for having supervised this work. I also thank my readers Tomaso Poggio, Victor Zue, and especially Mark Liberman. I am most grateful to Patrick Winston for his support at the M.I.T. Artificial Intelligence Laboratory and Osamu Fujimura, Max Matthews and James Flanagan for their support at Bell Laboratories.

SPEECH TIME-FREQUENCY REPRESENTATIONS

1

Introduction

In order to perceive speech and other sounds, the incoming sound wave must be transformed into a variety of representations, each bringing forth different aspects of the signal, its source, and meaning. Understanding how we perceive and how machines can be made to perceive auditory signals means, in part, discovering appropriate representations for the signals and how to compute them. For many kinds of sounds, little is known in this respect. What auditory features, for example, will distinguish a knock at the door from a footstep?

For speech signals, more is thought to be known. A phonetician will tell you, for example, that the /æ/ in *bad* can be distinguished from the /i/ in *bead* by the location of characteristic peaks in their respective spectra. He could even train you to identify a wide variety of phonetic elements by looking at their spectrograms. Formalizing this knowledge, however, so that a computer can do this well (in a general setting) has proved hard.

An analogy may explain why. I could train you to distinguish a Mercedes from some other car easily; I would just describe the hood ornament. † To train a machine to do this task would be much harder. Not only would I have to describe the hood ornament, but I would also have to provide all the visual abilities that I take for granted with a human — finding edges and boundaries, recognizing closed forms, etc. I believe the failure to correctly provide the corresponding auditory abilities — finding spectral "peaks" and temporal discontinuites, recognizing continuous forms, etc. — is an important reason why the speech recognition problem has been so difficult.

This problem is in some ways even harder than visual analysis. In vision, it is clear that the two-dimensional image is a natural starting point. In audition, a similar 2D representation is important, with time along one axis and frequency along the other. But how should this idea be made precise (the well-known uncertainty principle of fourier analysis is one of the thorny issues involved)? Should we use the conventional spectrogram, the Wigner distribution, a pseudo-auditory spectrogram, or something entirely new, and how should this decision be made?

In vision, the notion of edges, lines, and so forth obviously are important features of an image. In audition, it is harder to decide what are the appropriate primitive elements. Can some symbolic description summarize the relevant features of a sound's time-frequency representation analogous to how a line drawing summarizes an image?

These questions about the early steps in auditory processing are the topic of this book. The emphasis will be on speech signals primarily because the intermediate goals to which the initial computations must aim are better understood. I believe, nevertheless, that many

---

† I thank Mark Liberman for this example.

of the auditory processing issues discussed here are also relevant for other kinds of sounds.

The topic as stated is still too broad. Speech and other signals are made up of many different kinds of components. For instance, speech has fairly smoothly changing vocalic regions that are quite different from the more discontinuous structure of consonantal regions. It is unlikely that the same initial representations will be appropriate for every kind of signal. The emphasis here will be on signals like those found in the more continuous, sonorant regions of speech.

In the sonorant regions, we find an apparent feature is local spectral energy concentrations that vary in center frequency with time. These peaks are due, in part, to the "resonances" of the vocal tract – the so-called *formants*. The formant locations (labelled F1,F2,... in order of increasing frequency) specify the general vowel quality, r-coloring and roundness, while the formant transitions between consonants and vowels play an important role in consonant identification [see e.g. Chiba & Kajiyama 1941; Fant 1960; Liberman, et al 1954; Ladefoged 1975]. A. Liberman, in fact, claims that "...the second formant transition...is probably the single most important carrier of linguistic information in the speech signal [Liberman, et al 1967]. Thus, restricting the discussion to these regions is by no means uninteresting.

The initial speech processing envisioned here has been divided into two steps. The first step, which produces a joint time-frequency representation of the signal energy, is explored in Chapter 2 and Chapter 3. The second step, which produces a symbolic representation that captures the acoustically relevant features present in the joint time-frequency energy representation, is explored in Chapter 4 (see Figure 1.1).

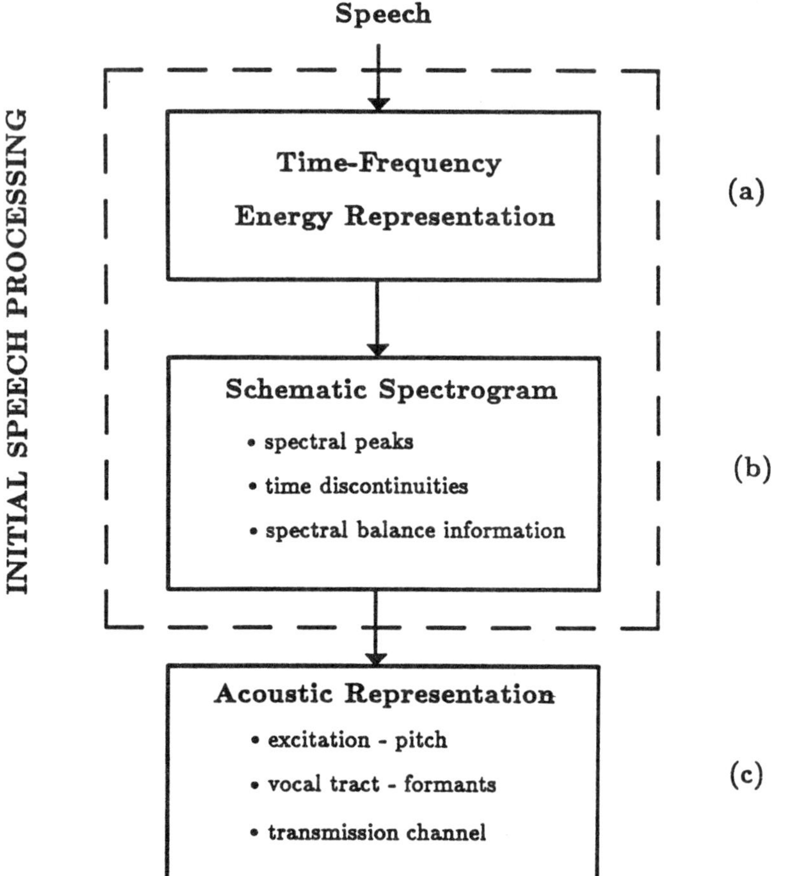

Figure 1.1. *The initial speech processing is seen as divided into two steps. (a) The first step represents the signal energy as joint functions of time and frequency. (b) The second step builds a symbolic representation of the significant features present in the joint time-frequency energy representations. At this step, which we call the* **schematic spectrogram,** *there is no undue commitment to the acoustic origin of the features represented; it is a description of the signal, not its sources. (c) In subsequent processing, these initial descriptions can be used to decompose the signal into its acoustic sources.*

One of the most difficult problems in deriving the form of such representations is deciding which properties or axioms to assume at the outset. If strong assumptions are made about the received signal, then rigorously defined optimal detection can result. For example, if we assume that the received signal consists solely of a known signal in additive Gaussian noise, then we could build a matched filter that performs optimal Bayesian detection [e.g., see Van Trees 1968]. The disadvantage of such strong assumptions is that they are seldom universally valid for natural perceptual signals.

On the other hand, weaker assumptions made about the received signal can be combined with assumptions about the design of the representation, things like linearity, continuity, locality, and stability, that can result in a solution [cf. Marr & Nishihara]. These design criteria are chosen not on the basis of a specific signal model, but instead as reasonable choices that should be appropriate for a wide range of natural signals. The disadvantage of this approach is that the justification of the design decisions is more intuitive and abstract.

In the best of circumstances, the two approaches would result in the same or similar solutions to a problem. Thus the auditory processing would perform optimally (in different senses) when both appropriate weak and strong assumptions are made about the received signal.

Chapter 2 derives those joint time-frequency energy representations that satisfy a small set of desirable properties; these properties are intentionally kept quite general. Chapter 3 re-examines this problem in a more specific setting. Given a (time-varying) model of speech production, what time-frequency representation of the signal best depicts the 'transfer function' of the vocal tract while suppressing the excitation. These two approaches, in fact, yield similar solutions.

In the initial part of Chapter 4, a general, heuristic argument is used to produce a phonetically relevant, symbolic representation of the signal. In a later part, these solutions are briefly related to a signal detection model.

In Chapter 5, we look at a wide range of examples using these proposed methods. We examine some traditionally difficult speech cases — glides and liquids, nasalized vowels, consonant-vowel transitions, female speech, and imperfect transmission channels.

N.B.: For the figures in this book, time is in seconds, frequency in hertz, and energy in decibels, unless otherwise indicated.

2

The Time-Frequency Energy Representation

This chapter explores the design of joint time-frequency energy representations for speech signals. A set of desirable properties for such representations to satisfy is proposed, and the relationships among these properties is discussed. This includes a general treatment of the 'uncertainty' relations that arise. The signal transforms that best satisfy these properties are then derived and examined.

2.1. The stationary case

We begin with an analysis of the special case of stationary signals. There is a large literature for this case; Rabiner & Schafer [1978] and Flanagan [1972] provide good reviews. The discussion of it here is very condensed and confined to topics that are relevant to the sequel.

A stationary signal is used here to roughly mean a signal whose frequency content does not vary with time. More precisely, we consider only determinstic signals that are periodic and random signals that are correlation-stationary. For both kinds of signals, the power spectrum, the fourier transform of the autocorrelation function, captures naturally the energy present at each frequency. † Time is removed from this representation; the power spectrum is a one-dimensional representation of energy as a function of frequency.

For speech signals there are, of course, no completely stationary signals. We can, however, deliberately utter vowels so that they are steady-state for as long as we like. Figure 2.1 shows the spectrum of a long duration, voiced /i/. We find in the spectrum many of the characteristic features of a steady-state vowel.

Let us examine the spectrum in Figure 2.1. Note the y-axis is logarithmic to compress the wide dynamic range of the speech. At a fine scale in this spectrum, there are peaks spaced about every hundred Hertz; these are the harmonics of the pitch. The somewhat larger scale peaks, of a few hundred Hz bandwidth, are the formant peaks. The peak at about 300 Hz is F1 and the peak at about 2300 Hz is F2, which is characteristic of an /i/ vowel for an adult male. Still larger scale shaping of the spectrum, so called *spectral balance*, is due to the formant locations, the nature of the voicing and the transmission channel.

The spectral structure of a vowel, therefore, is due acoustically to several factors: (1) the vocal excitation — e.g., voiced; (2) the vocal tract transfer function, characterized by its resonant frequencies — the formants, and (3) the transmission characteristics — e.g., room acoustics. Determining these factors from the speech (i.e., finding the formant frequencies, the pitch, etc.) is an important interme-

---

† For a deterministic signal $x(t)$, its autocorrelation function is $\int x(t + \tau)x^*(t)\, dt$, and for a stationary random process $y(t)$, its autocorrelation function is $\mathcal{E}[y(t + \tau)y^*(t)]$.

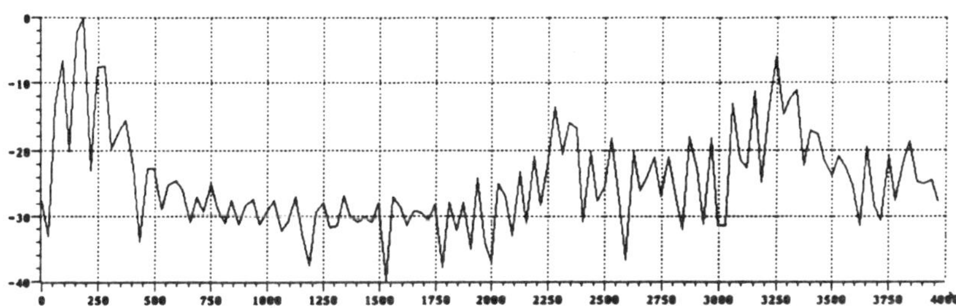

Figure 2.1. *Short-time log spectrum of a steady-state /i/. The finest scale structure corresponds to the harmonics of the pitch, spaced about every 100 Hz. At an intermediate scale are the formant peaks; e.g., F1 at 300 Hz and F2 at 2300 Hz. At the largest scale is the overall* **spectral balance.**

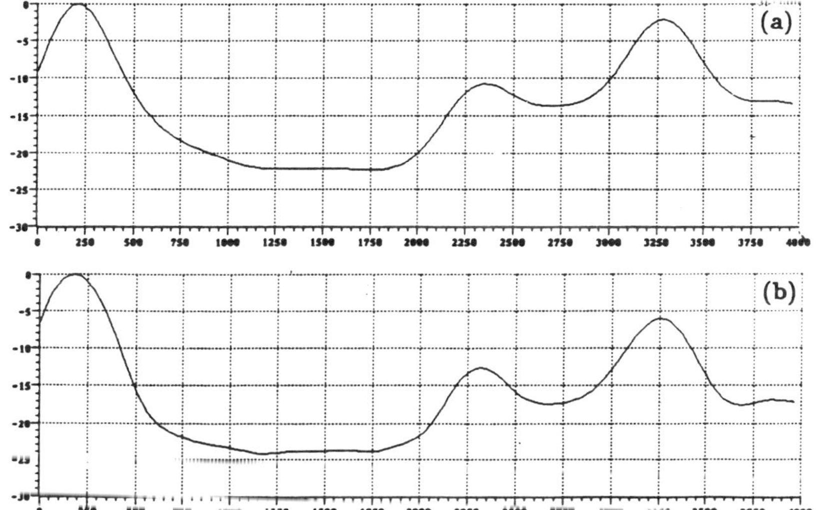

Figure 2.2. *Spectrum in Figure 2.1 smoothed to suppress the excitation. (a) Log spectrum convolved with gaussian (cepstral smoothing). (b) Power spectrum convolved with gaussian (and then transformed to a log scale).*

diate step in speech analysis, since they decompose the signal into components of nearly independent origin, and are (thus) starting points for the phonetician's description of speech signal.

A key point in separating these factors in the speech signal is that they operate at somewhat different scales in its spectrum; the fine scale structure is due mostly to the excitation, while the intermediate scale structure is due to the vocal tract transfer function. A common technique for selecting a scale of interest is to smooth the spectrum by linear convolution, or equivalently, to window the fourier transform of the spectrum. The fourier transform of the log spectrum is called the *cepstrum*, its dimension *quefrencies*, and the smoothing performed *cepstral smoothing* or *liftering*. [Oppenheim 1969; Oppenheim & Shafer 1975]. Figure 2.2a shows the spectrum in Figure 2.1 after it has been cepstrally smoothed at a scale to emphasize the formants, and suppress the excitation. We shall see in Chapter 3 that this operation, in fact, effectively separates excitation from transfer function in certain idealized, stationary cases.

It is smoothing the power spectrum, not its logarithm, that most easily generalizes to the non-stationary case later. We will therefore select our scales of interest by smoothing the power spectrum instead, or equivalently, by windowing its fourier transform, the autocorrelation function. Figure 2.2b shows the spectrum in Figure 2.1 after it has been thus smoothed. †

What should the form of the convolution kernel in this smoothing operation be? A desirable smoothing kernel would have good locality (or resolution) for a given amount of smoothing. In other words, it would have small duration for the given duration of its transform. These two durations are related by the uncertainty prin-

---

† Empirically, power and log smoothing often produce similar results.

ciple: given a function $h(x)$ with fourier transform $H(s)$, if the variance of $|h(x)|^2$ is $(\Delta x)^2$ and the variance of $|H(s)|^2$ is $(\Delta s)^2$, then $\Delta x \, \Delta s \geq \frac{1}{2}$ [Bracewell 1978]. Marr & Hildreth [1980] proposed a gaussian smoothing kernel (in a vision task) because it is the unique shape that meets the uncertainty principle with equality.

2.2. The quasi-stationary case

The previous section examined the analysis of stationary speech signals. No real speech signal, of course, is purely stationary. If the frequency content of a signal varies slowly with time, however, there is a simple extension of the previous results. The idea is to examine the signal over a short duration window. Given a signal $x(t)$ and a window $g(t)$, the *short-time* power spectrum at time t is

$$S_x(t,\omega) = \left| \int_{-\infty}^{\infty} g(\tau)x(t+\tau)e^{-i\omega\tau} \, d\tau \right|^2. \qquad (2.2.1)$$

Considered as a two-dimensional function of time and frequency, this signal representation is called a *spectrogram*. Many different window shapes have been used; they typically are symmetric, unimodal, and smooth, e.g., a gaussian or a raised single period of a cosine.

Signals for which a window can be found whose duration is long enough to allow adequate frequency resolution, but short enough to allow adequate time resolution are called *quasi-stationary*. The example of the previous section was, in fact, a quasi-stationary vowel. Virtually all speech analysis methods in the past depend on the quasi-stationary assumption.

2.3. Non-stationarity

There do exist signals for which no window duration is adequate. A very simple such signal is the linear *chirp*, $e^{i\frac{1}{2}mt^2}$, whose instantaneous frequency increases linearly with time. The quasi-stationary assumption breaks down for sufficently large modulation slope m of the signal. Let us examine this claim.

By the uncertainty principle, the product of the time duration Δt and the frequency duration (*bandwidth*) $\Delta\omega$ of a window is bounded below by 1/2. The window duration and bandwidth, in turn, determine the time and frequency resolution, respectively, in the short-time spectra. † In other words, if the window duration is too small, then the frequency resolution will be poor and if the window duration is too long, the time resolution will be poor. Further, for a non-stationary signal, poor time resolution can also mean poor frequency resolution since the frequency content will have changed over the duration of the window, blurring the spectrum.

To illustrate these points, consider the short-time spectrum of a linear chirp, $e^{i\frac{1}{2}mt^2}$, using a gaussian window, $e^{-t^2/2\sigma^2}$. We can measure the the relative bandwidth of the spectrum for different window sizes (σ's) in terms of the standard deviation of the spectrum ($\approx.42$ the half-power bandwidth), which is $\sqrt{(m^2\sigma^4 + 1)/2\sigma^2}$, where the units are seconds and radians. Note that when $m \neq 0$, this grows without bound as the window size becomes very small or very large. It has a minimum value of $\sqrt{m}$, which occurs when the standard deviation of the gaussian is $1/\sqrt{m}$.

We see from this that the minimum possible bandwidth of the short-time spectrum of a chirp (using a gaussian window) grows with increasing modulation slope. Figure 2.3 shows the short-time spectra of chirps of various modulation slopes using windows that give the minimum bandwidth. For a slope of 50 Hz/msec, the chirp peak has

---

† This is made precise by Theorem D in Section 2.6.

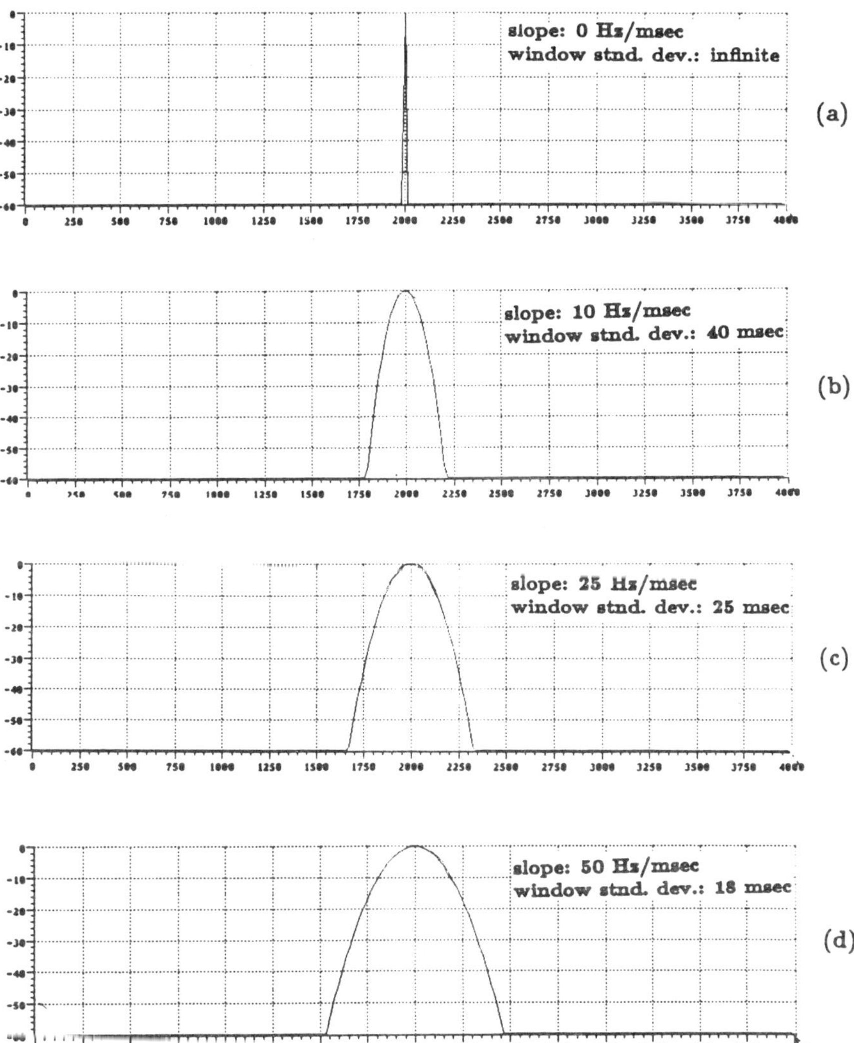

Figure 2.3. *Short-time spectra of linear chirps of several modulation slopes using gaussian windows that give the minimum bandwidth. At the largest slope, the chirp peak is significantly broadened*

been broadened by several hundred Hz in the spectrum. The point here is that, in theory, the usual quasi-stationary spectral analysis methods will give poor resolution for sufficiently non-stationary signals. A few examples from natural speech will show that such conditions arise in practice.

Figure 2.4 shows cepstrally smoothed, short-time spectra of various /w/'s, uttered first slowly and then increasingly rapidly. The spectrogram window used was a gaussian of 4 msec standard deviation, which has an effective duration of about a pitch period, the minimum duration that gives a reasonably stable spectral estimate. The cepstral window is also chosen as brief as possible, while still removing the harmonic peaks. Notice that the peak in the spectrum at about 1500 Hz, corresponding to F2, grows in bandwidth with the increasing slope of F2 as seen in the corresponding spectrograms in Figure 2.5. In case (c), where the F2 slope is about 40 Hz/msec, F2 is so broadened that its peak (i.e., the local maximum) is lost in the short-time spectrum. Such an F2 slope is not uncommon for a /w/. In /j/'s, F2 can have large negative slopes, and in /r/ contexts, F3 can have very steep slopes; see Figure 2.6. At consonant-vowel transistions, where the formant trajectories are considered very important for stop consonant identification [Liberman, et al 1954], the formant motion can also be very rapid; again see Figure 2.6.

It is worth noting that natural sounds other than the human voice can produce non-stationary signals that are "chirped." For instance, bird song and bat cries contain many rapid FM chirps [Greenewalt 1968; Marler 1979; Neuweiler 1977]. If a sound source is in relative motion to the listener then Doppler effects can cause large frequency shifts in the received signal across time [e.g., Dudgeon 1984]. † Glissandi

---

† Some bats (the so-called CF bats) emit continuous tones, evidently depending on Doppler shifts for echolocation.

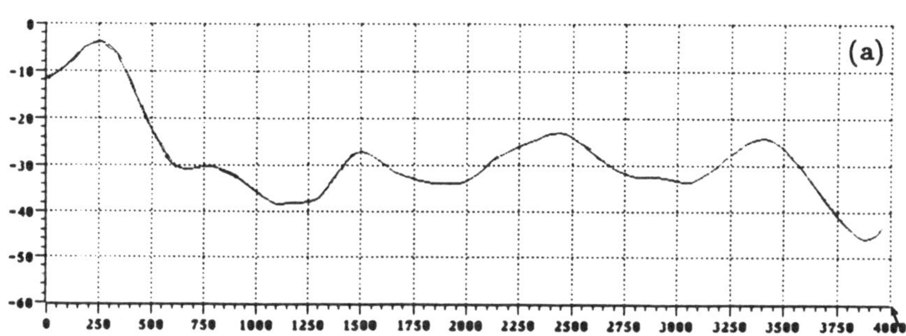

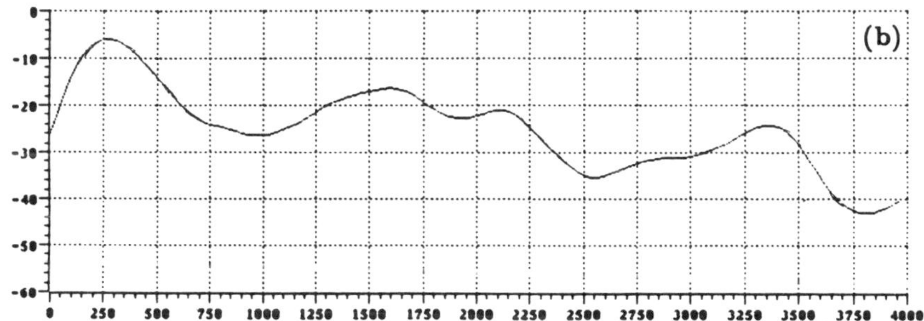

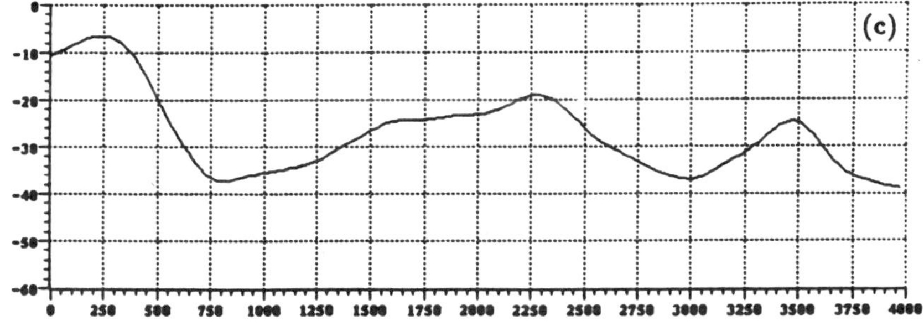

Figure 2.4. *Cepstrally smoothed, short-time spectra of /w/'s, uttered first very slowly, then increasingly rapidly. In (c), F2 is so broadened by the analysis that its peak (i.e., the local maximum) disappears. Cf. Figure 2.5.*

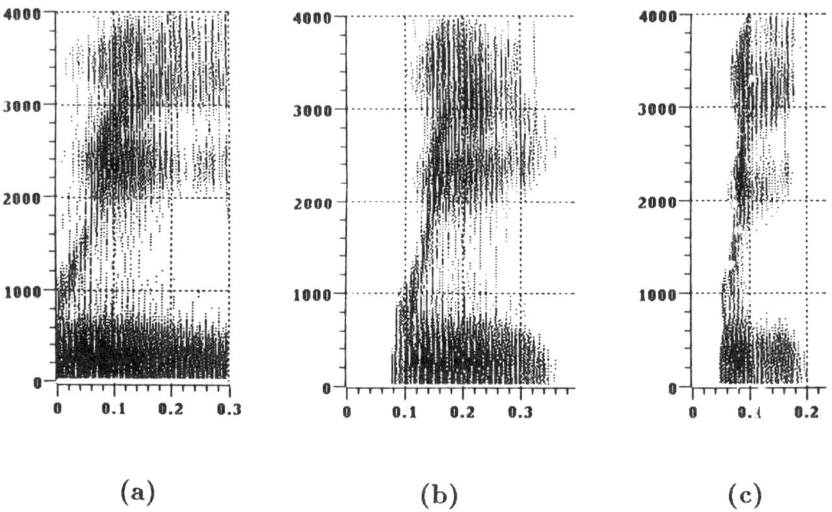

(a)   (b)   (c)

Figure 2.5. *Wide-band spectrograms of the /w/'s used in Figure 2.4. Note that F2 remains clearly visible with increasing slope in the two-dimensional display.*

of various musical instuments provide still more examples of signals that contain rapidly time-varying spectral content.

It is also suggestive that neurophysiologists have found that a large population of the auditory cells in the mammalian cochear nucleus do not respond optimally to continuous tones, but instead to sweep

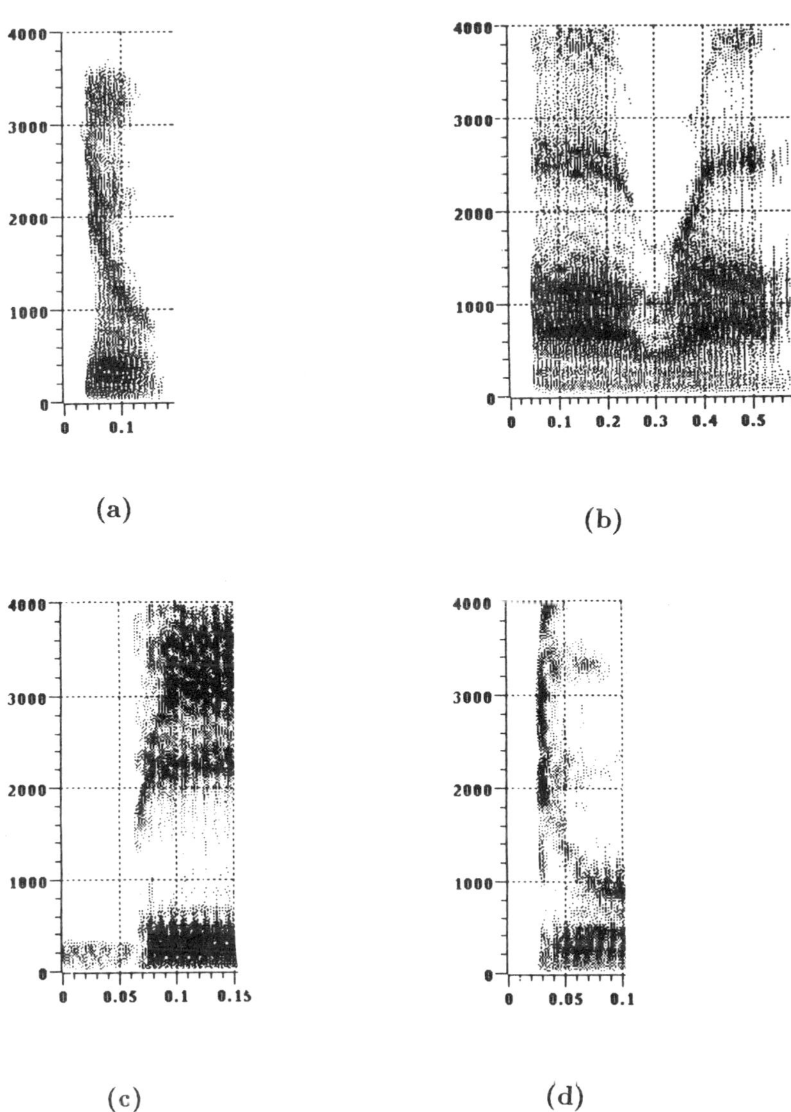

Figure 2.6. Spectrograms of rapid formant motion in various contexts. (a) /ju/. (b) /ara/. (c) /bi/ in the context /tubi/. (d) /du/ in the context /tldw/.

tones, with different populations responding to different preferred modulation slopes ranging over ±15 Hz/msec [Møller 1978; Britt & Starr 1976]. Further, psychophysical adaptation studies have shown similar directional selectivity in the human auditory system [Kay & Matthews 1972; Regan & Tansley 1979].

The above comments are meant to call into question the validity of the quasi-stationary assumption for speech and other auditory signals. We have seen that speech is not always quasi-stationary, even in the sonorant regions. Assuming so, means that important features will be missed, having been blurred by the analysis. It is interesting to note that while the individual short-time spectra of the non-stationary signals described above give a poor description of the signals, their spectrograms are nevertheless quite legible. This is because when we look at a spectrogram, we are not confined to examining them one-dimensionally along single frequency slices, but instead we see a two-dimensional time and frequency surface. In other words, time is not used as a parameter that varies over a family of spectra, but as one of the intrinsic dimensions of the representation.

I believe, in fact, that thinking of the initial speech processing as consisting of a family of independent one-dimensional spectral analyses parameterized by time is inappropiate. The problem should be thought of as a joint time-frequency analysis, with the relationships and trade-offs between the two dimensions directly addressed, which brings us to the next section.

2.4. Joint time-frequency representations

Various ways have been used to express signal energy as a joint function of time and frequency. Certainly the most popular is the spectrogram,

$$S_x(t,\omega) = \left| \int\limits_{-\infty}^{\infty} g(\tau)x(t+\tau)e^{-i\omega\tau}\, d\tau \right|^2, \qquad (2.4.1)$$

which is just the short-time spectra described above displayed two-dimensionally. The fact that the simultaneous time and frequency resolution in the spectrogram is bounded by the uncertainty relation has led others to seek representations that do not have this limitation.

This is usually formulated in terms of the *marginals* (or *projections*) of the signal representation $F_x(t,\omega)$ [Cohen 1966]. Let

$$\pi_1(t) = \frac{1}{2\pi} \int\limits_{-\infty}^{\infty} F_x(t,\omega)\, d\omega, \qquad (2.4.2a)$$

$$\pi_2(\omega) = \int\limits_{-\infty}^{\infty} F_x(t,\omega)\, dt. \qquad (2.4.2b)$$

Perfect time and frequency resolution in this formulation requires that

$$\pi_1(t) = |x(t)|^2 \qquad \text{and} \qquad \pi_2(\omega) = |X(\omega)|^2. \qquad (2.4.3)$$

An example of a joint time-frequency representation that satisfies these requirements is the Wigner distribution,

$$W_x(t,\omega) = \int\limits_{-\infty}^{\infty} e^{-i\omega\tau} x(t+\tau/2)x^*(t-\tau/2)\, d\tau, \qquad (2.4.4)$$

which is currently quite popular in the signal processing literature [Classen & Mecklenbräuker 1980a,c].

The Wigner distribution of an impulse, $x(t) = \delta(t-t_0)$ is $W_x(t,\omega) = \delta(t-t_0)$, i.e., the signal energy is taken to lie on the vertical line $t = t_0$

in the time-frequency plane. Similarly, for a complex exponential, $y(t) = e^{i\omega_0 t}$, the signal energy lies on the horizontal line at $\omega = \omega_0$ ($W_y(t,\omega) = 2\pi\delta(\omega - \omega_0)$), and for a linear chirp, $z(t) = e^{i(\omega_0 t + \frac{1}{2}mt^2)}$, the energy lies on the slanted line $\omega = mt + \omega_0$ ($W_z(t,\omega) = 2\pi\delta(\omega - \omega_0 - mt)$) (see Figure 2.7a).

In contrast, the spectrogram of these signals consist of broadened lines (see Figure 2.7b). There is, in fact, a simple relation between the spectrogram and the Wigner distribution of a signal $x(t)$:

$$S_x(t,\omega) = \frac{1}{2\pi} W_g(t,\omega) ** W_x(t,\omega), \qquad (2.4.5)$$

where $**$ denotes two-dimensional convolution and W_g is the Wigner distribution of the window [Classen & Mecklenbräuker 1980c]. If $g(t)$ is a gaussian, $\frac{1}{\sqrt{2\pi}\sigma}e^{-t^2/2\sigma^2}$, then its Wigner distribution is also simple; it is just a two-dimensional gaussian, $W_g(t,\omega) = \frac{1}{\sqrt{\pi}\sigma}e^{-t^2/\sigma^2}e^{-\omega^2\sigma^2}$. Thus, the two-dimensional convolution of the Wigner distributions in Figure 2.7a by a two-dimensional gaussian will give the spectrograms in Figure 2.7b.

If the duration of the gaussian spectrogram window is decreased, then the 2-D gaussian that, in essense, convolves the Wigner distribution to give the spectrogram becomes narrower in time, but wider in frequency, and vice versa. It should be clear from this example that the spectrogram does not meet the marginal requirement.

On the other hand, the Wigner distribution itself has some undesirable properties. In particular, multi-component signals give rise to cross terms that cannot be attributed much physical significance. For example, the Wigner distribution of $x(t) = \cos\omega_0 t$ is $W_x(t,\omega) = \frac{\pi}{2}[\delta(\omega - \omega_0) + \delta(\omega + \omega_0) + \delta(w)2\cos 2\omega_0 t]$ (see Figure 2.8a). The last term, which lies on a horizontal line at the frequency origin (varying sinusoidily in amplitude), seems spurious.

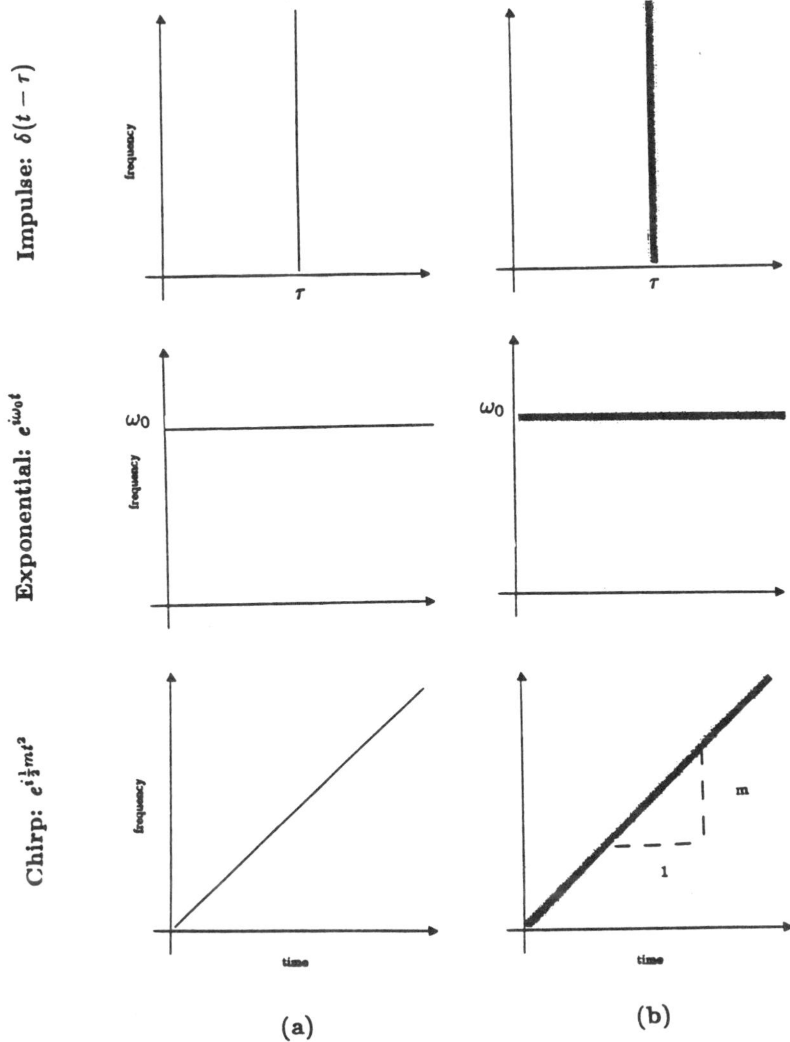

Figure 2.7. *Wigner distribution and spectrogram of some mono-component signals. (a) The Wigner distribution resolves these signals as perfectly narrow lines in the time-frequency plane. (b) The spectrogram is a smoothed version of the Wigner distribution (e.g., if the spectrogram window is a gaussian, then the smoothing kernel is a 2-D gaussian). The lines are broadened in this representation.*

The spectrogram of $cos\,\omega_0 t$, however, is just two broadened lines at $\omega = \pm\omega_0$, which seems better behaved with respect to superposition, since $cos\,\omega_0 t = \frac{1}{2}(e^{i\omega_0 t} + e^{-i\omega_0 t})$ (see Figure 2.8b). The cross term is, in effect, smoothed out by the convolution that transforms the Wigner distribution into the spectrogram.

These examples illustrate that there are various (possibly conflicting) properties that we might desire of a time-frequency representation, e.g., good time and frequency resolution, and superposition for multi-component signals. We shall, in fact, approach the problem of choosing our time-frequency energy representation by first specifying a set of desirable properties that the transform should satisfy, and then deriving its form.

2.5. Design criteria for time-frequency representations

We will restrict the discussion to the quadratic transforms of the signal, which have the form

$$F_x(t,\omega) = \int\limits_{-\infty}^{\infty} \int\limits_{-\infty}^{\infty} h(\tau_1, \tau_2; t, \omega) x(\tau_1) x^*(\tau_2) \, d\tau_1 \, d\tau_2, \qquad (2.5.1)$$

where $h(\tau_1, \tau_2; t, \omega)$ is an arbitrary function. This condition is imposed because it results in a particularly manageable class, and because the representation of energy as a quadratic function of the signal seems reasonable by analogy to other definitions of energy. The class is quite large and includes many of the joint time-frequency representations that have been previously proposed, such as the spectrograms, the Wigner distribution, and the Rihaczek distribution [cf. Claasen & Mecklenbräuker 1980c].

From this class of representations, we seek ones that satisfy the following criteria:

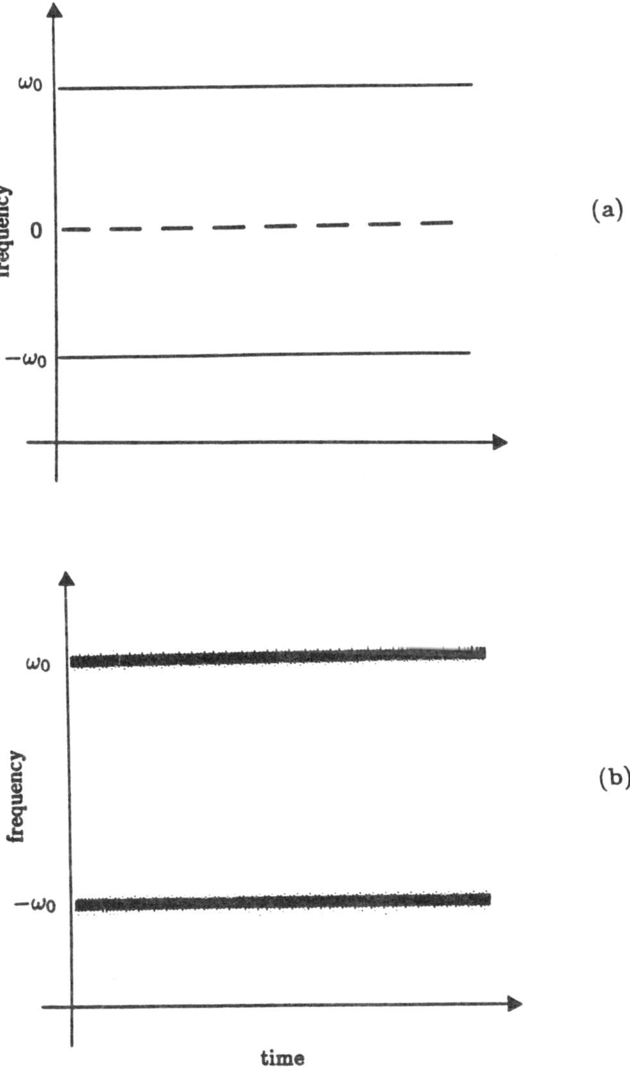

Figure 2.8. *Wigner distribution and spectrogram for* $\cos \omega_0 t$*. (a) The Wigner distribution of this signal has the 'spurious' cross term* $\delta(\omega)2\cos 2\omega_0 t$ *at the origin. (b) The spectrogram does not show this term; it has been, in effect, smoothed out.*

(C1) Shift invariance: A shift in time or frequency of the signal should result in a corresponding shift in time or frequency in the transform. Let $y(t) = x(t - \tau)$ and $z(t) = e^{i\varphi t}x(t)$. Then we require $F_y(t,\omega) = F_x(t - \tau, \omega)$ and $F_z(t,\omega) = F_x(t, \omega - \varphi)$. This property is desirable if we want to interpret the two dimensions of the transform as time and frequency.

Transforms satisfying this condition can be put in the forms

$$F_x(t,\omega) = \frac{1}{2\pi}\phi(t,\omega) ** W_x(t,\omega) \qquad (2.5.2)$$

and

$$F_x(t,\omega) = \mathcal{F}^{-1}[\Phi(\tau,\nu)A_x(\tau,\nu)], \qquad (2.5.3)$$

where "$**$" denotes two-dimensional convolution, W_x is the Wigner distribution

$$W_x(t,\omega) = \int\limits_{-\infty}^{\infty} e^{-i\omega\tau}x(t + \tau/2)x^*(t - \tau/2)\,d\tau, \qquad (2.5.4)$$

$\phi(t,\omega)$ is an arbitrary kernel function, $\mathcal{F}$ is the 2-D fourier transform in the form $\mathcal{F}[q(t,\omega)] = \frac{1}{2\pi}\int\limits_{-\infty}^{\infty}\int\limits_{-\infty}^{\infty} e^{i(-\nu t + \tau\omega)}q(t,\omega)\,dt\,d\omega$, $\Phi(\tau,\nu) = \mathcal{F}[\phi(t,\omega)]$, and A_x is the time-frequency autocorrelation function †

$$A_x(\tau,\nu) = \mathcal{F}[W_x(t,\omega)] = \int\limits_{-\infty}^{\infty} e^{-i\nu t}x(t + \tau/2)x^*(t - \tau/2)\,dt \quad (2.5.5)$$

for $x(t)$ [Claasen & Mecklenbräuker 1980c]. Note that for a spectrogram, $\phi(t,\omega)$ is the Wigner distribution of the spectrogram window, by Eq. 2.4.5 and Eq. 2.5.2.

---

† Some authors call this the *ambiguity function* [e.g., Claasen & Mecklenbräuker 1980a]; others reserve this term for $|A_x(\tau,\nu)|^2$ [e.g., Van Trees 1968].

(C2) Positivity: The signal energy at a given point in time and frequency should be real and positive: $F_x(t,\omega) \geq 0$ for all x, t, and ω. This seems appropriate for interpreting the transform as an energy distribution. Some authors have argued against the positivity requirement [e.g. Claasen & Mecklenbräuker 1980c]. We shall examine the consequences of lifting this condition in the next section.

(C3) Superposition: This idea is that the time-frequency representation of a multi-component signal should be a simple superposition of its components. The straight-forward linear formulation of this, i.e., $F_{x+cy}(t,\omega) = F_x(t,\omega) + cF_y(t,\omega)$, however, is inconsistent with the quadratic nature of the transform, and the shift-invariance property **C1**. This apparent shortcoming is also true, for example, of the spectrogram (Eq. 2.4.1). Nevertheless, we usually think of the conventional spectrogram as being well-behaved under superposition. This is because *non-overlapping* components do superimpose, i.e., $S_{x+y}(t,\omega) = S_x(t,\omega) + S_y(t,\omega)$ when $S_x(t,\omega)S_y(t,\omega) = 0$. There are no cross terms in this case. On the other hand, the Wigner distribution does not have this property, suffering from cross terms to which there cannot be attributed much physical significance.

We shall require this property for our time-frequency representation, namely

$$F_{x+y}(t,\omega) = F_x(t,\omega) + F_y(t,\omega) \quad \text{when} \quad F_x(t,\omega)F_y(t,\omega) = 0.$$
$$(2.5.6a)$$

More generally, we would like $F_{x+y}(t,\omega) \approx F_x(t,\omega) + F_y(t,\omega)$ when $F_x(t,\omega)F_y(t,\omega) \approx 0$. Stated more precisely, we require for any $\epsilon > 0$, there exists a $\delta > 0$ such that

$$|F_{x+y}(t,\omega) - [F_x(t,\omega) + F_y(t,\omega)]| < \epsilon \quad \text{when} \quad |F_x(t,\omega)F_y(t,\omega)| < \delta.$$
$$(2.5.6b)$$

(C4) Locality: Signal energy that is localized in time-frequency should remain localized in time-frequency in the transform. The advantage of the Wigner distribution is that it is perfectly localized according to various criteria, such as preserving the marginal distributions (Eq. 2.4.3) and the finite support properties [see Claasen & Mecklenbräuker 1980a]. † The Wigner distribution, however, does not satisfy the positivity (**C2**) or superposition (**C3**) properties, as indicated earlier. In fact, positivity (and thus, as we shall see, superposition) is inconsistent with the time and frequency marginal conditions [Claasen & Mecklenbräuker 1980c]. Fortunately, for our purposes, we do not require perfect locality, so we can relax the above conditions somewhat.

From Eq. 2.5.2, the transform kernel $\phi(t,\omega)$ can be viewed as the point spread function on the perfectly localized Wigner distribution. We can therefore measure the locality of the transform in time and frequency in terms of the variances ‡

$$\sigma_t^2 = \frac{\int \int t^2 |\phi(t,\omega)|^2 \, dt \, d\omega}{\int \int |\phi(t,\omega)|^2 \, dt \, d\omega}, \qquad (2.5.7a)$$

---

† The finite support property states that if a signal has finite extent in time or frequency then its representation will have the same extent in the corresponding variable.

‡ The generality of this approach depends on the Wigner distribution *uniquely* satisfying 'perfect' locality. Cohen has shown that a quadratic transform that satisfies the shift-invariance property (**C1**) will meet the time and frequency marginal conditions (Eq. 2.4.3) if $\Phi(\tau,0) = 1$ for all τ and $\Phi(0,\nu) = 1$ for all ν. These marginal conditions essentially guarantee that an impulse and a complex exponential are not 'blurred' by the time-frequency representation, but are not strong enough to also guarantee that a linear chirp is not 'blurred' (see Figure 2.7a). This additional condition is met uniquely by the Wigner distribution. In other words, we interpret perfect locality to mean that the signal transform does not spread the signal energy in *any* direction in time-frequency (not just the horizontal and vertical directions). We postpone a more thorough discussion of this point until Section 2.8, when the necessary mathematical machinery will be introduced.

and

$$\sigma_\omega^2 = \frac{\int\int \omega^2 |\phi(t,\omega)|^2 \, dt \, d\omega}{\int\int |\phi(t,\omega)|^2 \, dt \, d\omega}, \qquad (2.5.7b)$$

where we assume that the center of mass of $|\phi(t,\omega)|^2$ is at the origin.
*

In general, these two measures are not enough; an additional locality measure is important, the covariance

$$\sigma_{t\omega} = \frac{\int\int t\omega |\phi(t,\omega)|^2 \, dt \, d\omega}{\int\int |\phi(t,\omega)|^2 \, dt \, d\omega}. \qquad (2.5.7c)$$

Together, σ_t, σ_ω, and $\sigma_{t\omega}$ determine the covariance matrix and the associated concentration ellipse in the (t,ω) plane,

$$\begin{pmatrix} t & \omega \end{pmatrix} \begin{pmatrix} \sigma_t^2 & \sigma_{t\omega} \\ \sigma_{t\omega} & \sigma_\omega^2 \end{pmatrix}^{-1} \begin{pmatrix} t \\ \omega \end{pmatrix} = 1. \qquad (2.5.8)$$

When $\sigma_{t\omega} = 0$, the major and minor axes of the concentration ellipse coincide with the time and frequency axes (Figure 2.9a). More generally, the concentration ellipse may be oriented obliquely relative to the co-ordinate axes (Figure 2.9b). We shall call transforms that satisfy the condition $\sigma_{t\omega} = 0$ on their kernel *non-directionally localized*. This name is appropriate since we can rescale the co-ordinate axes to make the concentration ellipse a circle under this condition. Thus viewed, the transform spreads energy uniformly in all directions in time-frequency. On the other hand, if $\sigma_{t\omega} \neq 0$, then this does not hold, and the transform will be *directionally localized*, always having better resolution in some time frequency directions than others regardless of the scaling of the axes.

---

* This assumption is not very restrictive on the form of the transform, since we can always shift $\phi(t,\omega)$ in time and frequency to satisfy it. This shift, in turn, shifts the transform in time and frequency.

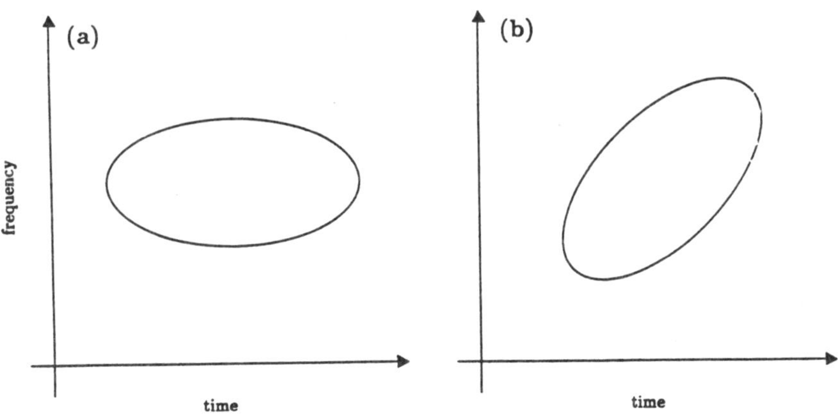

Figure 2.9. *Concentration ellipses for transform kernels. (a) Non-directional kernel ($\sigma_{t\omega} = 0$): the co-ordinate axes can be re-scaled to make the concentration ellipse a circle. Thus viewed, the corresponding transform spreads the signal energy equally in all time-frequency directions. (b) Directional kernel ($\sigma_{t\omega} \neq 0$): the co-ordinate axes cannot be re-scaled to make the concentration ellipse a circle. The corresponding transform always has better resolution in some time-frequency directions than others.*

The analysis of the non-directional transforms is more straightforward. We therefore restrict our attention to this case until Section 2.8, when we shall examine the more general case. We will see there that the principal results are essentially the same as non-directional case, suitably generalized. The analysis, however, is more complex, and is thus best left until later.

To summarize, given a non-directional transform ($\sigma_{t\omega} = 0$), σ_t and σ_ω measure its degree of locality in time and frequency. The smaller

σ_t and σ_ω are, the better the time and frequency resolution.

(C5) Smoothness: Similar to the stationary case, different aspects of the speech signal can arise at different scales in time-frequency. For example, voiced excitation can give rise to fine scale structure on the order of the pitch period in the time dimension and the fundamental frequency in the frequency dimension. The formant structure, on the other hand, arises at a somewhat larger scale. Thus, one of the design parameters for our transform is the scale in time-frequency we wish to examine. Said differently, we want the transform to be smooth in time-frequency to a given degree.

This notion of scale can be be formalized by measuring the distribution of the spatial frequencies present in $F_x(t,\omega)$, i.e., the distribution of energy about the origin of its 2-D fourier transform. Since $\mathcal{F}[F_x(t,\omega)] = \Phi(\tau,\nu)A_x(\tau,\nu)$ (Eq. 2.5.3), the relative amount of spread is determined by the choice of $\Phi(\tau,\nu)$, which windows the time-frequency autocorrelation function. We can measure this spread in terms of the variances

$$\Sigma_\tau^2 = \frac{\int\int \tau^2 |\Phi(\tau,\nu)|^2 \, d\tau \, d\nu}{\int\int |\Phi(\tau,\nu)|^2 \, d\tau \, d\nu} \tag{2.5.9a}$$

$$\Sigma_\nu^2 = \frac{\int\int \nu^2 |\Phi(\tau,\nu)|^2 \, d\tau \, d\nu}{\int\int |\Phi(\tau,\nu)|^2 \, d\tau \, d\nu}, \tag{2.5.9b}$$

and

$$\Sigma_{\tau\nu} = \frac{\int\int \tau\nu |\Phi(\tau,\nu)|^2 \, d\tau \, d\nu}{\int\int |\Phi(\tau,\nu)|^2 \, d\tau \, d\nu}, \tag{2.5.9c}$$

where we assume that the center of mass of $|\Phi(\tau,\nu)|^2$ is at the origin. † These determine the covariance matrix and the associated concentration ellipse in the (τ,ν) plane,

$$(\tau \quad \nu) \begin{pmatrix} \Sigma_\tau^2 & \Sigma_{\tau\nu} \\ \Sigma_{\tau\nu} & \Sigma_\nu^2 \end{pmatrix}^{-1} \begin{pmatrix} \tau \\ \nu \end{pmatrix} = 1. \tag{2.5.10}$$

---

† This assumption will be true if the transform is real.

When $\Sigma_{\tau\nu} = 0$, we call the transform *non-directionally smooth*. In this case, it is possible to rescale the co-ordinate axes to make the concentration ellipse a circle, and thus viewed the transform smoothes the signal in time-frequency uniformly in all direction in time-frequency. On the other hand, if $\Sigma_{\tau\nu} \neq 0$, then this does not hold, and the transform will be *directionally smooth*, always smoothing more in some time-frequency directions than others regardless of the scaling of the axes. Just like the locality condition, we will restrict attention now to the non-directional transforms. We consider the more general case in Section 2.8.

To summarize, given a non-directional transform $(\Sigma_{\tau\nu=0})$, Σ_τ and Σ_ν measure its scale in time and frequency. The smaller Σ_τ and Σ_ν are, the larger the selected scales.

Observe at this point the parallels between the stationary and non-stationary analyses. If we think of the Wigner distribution as the non-stationary analog to the raw power spectrum, then the time-frequency autocorrelation function (the Wigner distribution's 2-D fourier transform) is the 2-D analog to the autocorrelation function (the power spectrum's fourier transform). Further, windowing the time-frequency autocorrelation function smoothes the Wigner distribution, just as windowing the autocorrelation smoothes the raw spectrum. In both cases, the design decisions for the resulting transform require selecting a convolution kernel that satisfies both locality and smoothness requirments. In fact, we shall see in the next chapter that the analogy is even closer.

2.6. Relations among the design criteria

The various design criteria for our time-frequency energy representation are not independent. We shall state the important relationships

among them in this section. Throughout this section we assume that
the input signal $x(t)$ is finite energy, (i.e., $x \epsilon \mathcal{L}_2$) and that $F_x(t,\omega)$ is a
quadratic transform of the signal. This means that $F_x(t,\omega) = \langle Tx, x \rangle$
where $\langle x, y \rangle = \int_{-\infty}^{\infty} x(\alpha)y^*(\alpha) \, d\alpha$ and $T_{t,\omega}$ is a (bounded) linear operator on $\mathcal{L}_2$.

• **Shift-invariance & Positivity:** Together these imply that the
transform can be expressed as a superposition of spectrograms. †

Theorem A. *Let $F_x(t,\omega)$ be positive and shift-invariant. Then it
has the form*

$$F_x(t,\omega) = \int_{-\infty}^{\infty} S_x(t,\omega; g_\alpha) \, d\alpha, \qquad (2.6.1)$$

where $S_x(t,\omega; g)$ is the spectrogram having g as its window.

Proof: The positivity of $F_x(t,\omega)$ means that $T_{t,\omega}$ is a positive operator and therefore has a square root A, i.e.,

$$F_x = \langle A^*Ax, x \rangle = \langle Ax, Ax \rangle = \|Ax\|^2, \qquad (A.1)$$

where $\|x(\alpha)\|^2 = \int |x(\alpha)|^2 \, d\alpha$ [see Rudin 1973]. Representing the
linear operator A in terms of its impulse response $A_{t,\omega}[x(\alpha)] = \int h(\tau, \alpha; t, \omega)x(\tau) \, d\tau$ and substituting into Eq. A.1 gives

$$F_x(t,\omega) = \int_{-\infty}^{\infty} \left| \int_{-\infty}^{\infty} h(\alpha, \tau; t, \omega)x(\tau) \, d\tau \right|^2 d\alpha. \qquad (A.2)$$

---

† Bouachache, et al [1979] incorrectly state that a positive and shift-invariant
quadratic transform is necessarily a spectrogram.   Claasen & Mecklenbräuker [1984] point out this error, mentioning that linear combinations
of spectrograms must be included.

By time and frequency shift-invariance,

$$F_x(t+a, \omega+\varphi) = \int\limits_{-\infty}^{\infty} \left| \int\limits_{-\infty}^{\infty} h(\alpha,\tau;t,\omega)x(\tau+a)e^{-i\varphi(\tau+a)}\,d\tau \right|^2 d\alpha.$$

Setting $t = \omega = 0$ gives

$$F_x(a,\varphi) = \int\limits_{-\infty}^{\infty} \left| \int\limits_{-\infty}^{\infty} h(\alpha,\tau;0,0)x(\tau+a)e^{-i\varphi(\tau+a)}\,d\tau \right|^2 d\alpha,$$

or, with $g_\alpha(\tau) = h(\alpha,\tau;0,0)$,

$$= \int\limits_{-\infty}^{\infty} \left| \int\limits_{-\infty}^{\infty} g_\alpha(\tau)x(\tau+a)e^{-i\varphi(\tau+a)}\,d\tau \right|^2 d\alpha.$$

From Eq. 2.4.1, we see the outer integrand is the spectrogram $S_x(t,\omega;g_\alpha)$, giving Eq. 2.6.1. ///

• **Positivity & Superposition:** The next theorem shows that positivity implies superposition. In fact, it implies a strong form of superposition, as in Eq. 2.5.6b.

Theorem B. *If $F_x(t,\omega)$ is positive, then*

$$|F_{x+y}(t,\omega) - [F_x(t,\omega) + F_y(t,\omega)]|^2 \leq 4F_x(t,\omega)F_y(t,\omega). \qquad (2.6.2)$$

Proof: From the elementary fact about inner products

$$\|p+q\|^2 = \|p\|^2 + 2\,Re\,\langle p,q\rangle + \|q\|^2$$

it follows that

$$\left| \|p+q\|^2 - [\|p\|^2 + \|q\|^2] \right|^2 = 4|Re\,\langle p,q\rangle|^2$$

$$\leq 4|\langle p,q\rangle|^2.$$

Since $\langle p,q\rangle \leq \|p\|\,\|q\|$,

$$\leq 4\|p\|^2\|q\|^2.$$

Substituting $p = Ax$ and $q = Ay$ above and using Eq. A.1 gives Eq. 2.6.2. ///

If the transform is real, the converse of this theorem is also true; i.e., superposition implies either F_x or $-F_x$ is positive.

Theorem C. *Let $F_x(t,\omega)$ be real and satisfy superposition (Eq. 2.5.6a). Then either $F_x(t,\omega)$ or $-F_x(t,\omega)$ is positive.*

Proof: Step 1. First we show under the hypotheses of the theorem $\langle Tx,x\rangle = 0 \Rightarrow Tx = 0$.

Superposition says

$$\langle Tx,x\rangle\langle Ty,y\rangle = 0 \Rightarrow \langle T(x+y),x+y\rangle = \langle Tx,x\rangle + \langle Ty,y\rangle. \quad (C.1)$$

Since the form $\langle Tx,x\rangle$ is always real, $\langle Tx,y\rangle = \langle Ty,x\rangle^*$, so

$$\langle T(x+y),x+y\rangle = \langle Tx,x\rangle + 2\,Re\langle Tx,y\rangle + \langle Ty,y\rangle.$$

Thus, from Eq. C.1,

$$\langle Tx,x\rangle\langle Ty,y\rangle = 0 \Rightarrow Re\langle Tx,y\rangle = 0 \quad\quad (C.2)$$

Substituting ix into Eq. C.2 shows that $Im\langle Tx,y\rangle = 0$ also, so that

$$\langle Tx,x\rangle\langle Ty,y\rangle = 0 \Rightarrow \langle Tx,y\rangle = 0 \quad\quad (C.3)$$

Suppose that $\langle Tx, x \rangle = 0$. Then by Eq. C.3, $\langle Tx, y \rangle = 0$ for all y. If we let $y = Tx$, then $\langle Tx, Tx \rangle = 0$ and thus $Tx = 0$, as desired.

Step 2. We now show that $\langle Tz, z \rangle = 0 \Rightarrow Tz = 0$ implies $\pm T$ is positive. Suppose $\langle Tx, x \rangle > 0$ and $\langle Ty, y \rangle < 0$. Let $z = kx + y$ where k is real. Then

$$\langle Tz, z \rangle = k^2 \langle Tx, x \rangle + 2k\, Re\langle Tx, y \rangle + \langle Ty, y \rangle.$$

This is a quadratic in k, and since $\langle Tx, x \rangle \langle Ty, y \rangle < 0$, it has two distinct real zeroes. However, since $Tx \neq 0$, $Tz = kTx + Ty$ has at most one zero in k. Therefore, there exists a value of k such that $\langle Tz, z \rangle = 0$ but $Tz \neq 0$, contradicting the hypothesis, and implying $\pm T$ is positive. $///$

This last theorem shows that we can replace the positivity condition (**C2**) with the sole requirement that the transform be always real, and have an equivalent set of properties. In other words, the transform will necessarily be positive if superposition holds, and if positivity is abandoned, cross terms will necessarily prove a problem for multi-component signals such as speech.

• **Positivity & Locality:** The positivity condition places a limit on the time-frequency locality of the transform. When the transform is positive, it is sometimes convenient to measure locality in terms of the variances of $\phi(t, \omega)$ instead of $|\phi(t, \omega)|^2$. We define

$$\sigma_T^2 = \frac{\int \int t^2 \phi(t, \omega)\, dt\, d\omega}{\int \int \phi(t, \omega)\, dt\, d\omega} \qquad (2.6.3a)$$

and

$$\sigma_\Omega^2 = \frac{\int \int \omega^2 \phi(t, \omega)\, dt\, d\omega}{\int \int \phi(t, \omega)\, dt\, d\omega}, \qquad (2.6.3b)$$

where we assume that the center of mass of $\phi(t,\omega)$ is at the origin.
† When the transform is positive, we claim that these variances are
non-negative. To show this, first suppose the transform is a spectrogram. Then $\phi(t,\omega)$ is the Wigner distribution of the spectogram
window $g(t)$, and using Eq. 2.4.3, it is easy to see that

$$\sigma_T^2 = \operatorname*{var}_t |g(t)|^2 \quad \text{and} \quad \sigma_\Omega = \operatorname*{var}_\omega |G(\omega)|^2, \tag{2.6.4}$$

which are clearly non-negative [cf. DeBruin]. More generally, if the
transform is positive, it follows directly from Theorem A that

$$\sigma_T^2 = \int c_\alpha \operatorname*{var}_t |g_\alpha(t)|^2 \, d\alpha \quad \text{and} \quad \sigma_\Omega^2 = \int c_\alpha \operatorname*{var}_\omega |G_\alpha(\omega)|^2 \, d\alpha \tag{2.6.5}$$

where

$$c_\alpha = \frac{\displaystyle\int_{-\infty}^{\infty} |g_\alpha(t)|^2 \, dt}{\displaystyle\int\!\!\int_{-\infty}^{\infty} |g_{\alpha'}(t)|^2 \, dt \, d\alpha'} \tag{2.6.6}.$$

These are again non-negative quantities.

Eq. 2.6.5 shows that σ_T^2 is the (weighted) average window variance
in the representation of $F_x(t,\omega)$ as a superposition of spectrograms.
Since a spectrogram's values at a given time depend only on signal
values under its window, we see that a positive transform at a time
t effectively depends only on signal values within a few σ_T of t. *

---

† This assumption is necessary for the term 'variance' to apply. It is not
necessary, however, for the uncertainty relations presented below to be true
[cf. DeBruin].

* This is a stronger notion of time locality than in the previous section.
There, time locality essentially measured how the transform spread an impulse. The Wigner distribution is perfectly localized in this sense, because
it represents the energy of an impulse at time t_0 entirely on the vertical line
$t = t_0$ in the time-frequency plane. This does not mean that the Wigner

The next theorem states an important uncertainty relation for positive transforms. It bounds the simultaneous time and frequency resolution that can be obtained by such a transform.

Theorem D. *Let $F_x(t,\omega)$ be positive and shift-invariant. Then $\sigma_T \sigma_\Omega \geq \frac{1}{2}$.*

Proof: From Eq. 2.6.5,

$$\sigma_T^2 \sigma_\Omega^2 = \int c_\alpha \sigma_\alpha^2 \, d\alpha \int c_\alpha \Sigma_\alpha^2 \, d\alpha$$

where $\sigma_\alpha^2 = \operatorname*{var}_t |g_\alpha(t)|^2$ and $\Sigma_\alpha^2 = \operatorname*{var}_\omega |G_\alpha(\omega)|^2$. By the Schwarz Inequality,

$$\sigma_T^2 \sigma_\Omega^2 \geq \left(\int c_\alpha \sigma_\alpha \Sigma_\alpha \, d\alpha \right)^2 .$$

The classical uncertainty relation applied to $g_\alpha(t)$ gives $\sigma_\alpha \Sigma_\alpha \geq \frac{1}{2}$, so

$$\sigma_T^2 \sigma_\Omega^2 \geq \left(\frac{1}{2} \int c_\alpha \sigma_\alpha \, d\alpha \right)^2 = \frac{1}{4},$$

since $\int c_\alpha \, d\alpha = 1$ from Eq. 2.6.6. Taking square roots yields the desired result. ///

$\bullet$ **Locality & Smoothness:** Just as in the stationary case, locality and smoothness are conflicting properties. Greater smoothness means poorer locality and vice versa, other things being equal. This

---

distribution's values at time t_0 depend only on the signal value at t_0. Quite the opposite is true, they depend on the entire signal. (In fact, the signal can be recovered from the Wigner distribution's values at any fixed time t_0 (up to a multiplicative constant) [see Claasen & Mecklenbräuker 1980a].) However, when the transform is positive these two notions of locality coincide.

follows formally from a two-dimensional generalization of the classical uncertainty relation.

Theorem E. *If $F_x(t,\omega)$ is shift-invariant, then $\sigma_\tau \Sigma_\nu \geq \frac{1}{2}$ and $\sigma_\omega \Sigma_\tau \geq \frac{1}{2}$, with equality in both these relations iff*

$$\phi(t,\omega) \propto e^{-t^2/2\sigma_T^2} e^{-\omega^2/2\sigma_\Omega^2}. \tag{2.6.7}$$

Proof: First, we show that $\sigma_t \Sigma_\nu \geq \frac{1}{2}$, Let $\lambda(t,\tau) = \frac{1}{2\pi} \int \phi(t,\omega) e^{i\omega\tau}\, d\omega$. Then $\Phi(\tau,\nu) = \mathcal{F}[\phi(t,\omega)] = \int \lambda(t,\tau) e^{-it\nu}\, dt$. Applying the classical uncertainty relation to $\lambda(t,\tau)$ w.r.t. t gives

$$\frac{1}{2} \left(\int_{-\infty}^{\infty} |\lambda(t,\tau)|^2\, dt \int_{-\infty}^{\infty} |\Phi(\tau,\nu)|^2\, d\nu \right)^{\frac{1}{2}}$$

$$\leq \left(\int_{-\infty}^{\infty} t^2 |\lambda(t,\tau)|^2\, dt \int_{-\infty}^{\infty} \nu^2 |\Phi(\tau,\nu)|^2\, d\nu \right)^{\frac{1}{2}} \tag{E.1}$$

Integrating E.1 over τ and using the Schwarz Inequality

$$\frac{1}{2} \int_{-\infty}^{\infty} \left(\int_{-\infty}^{\infty} |\lambda(t,\tau)|^2\, dt \int_{-\infty}^{\infty} |\Phi(\tau,\nu)|^2\, d\nu \right)^{\frac{1}{2}} d\tau$$

$$\leq \int_{-\infty}^{\infty} \left(\int_{-\infty}^{\infty} t^2 |\lambda(t,\tau)|^2\, dt \int_{-\infty}^{\infty} \nu^2 |\Phi(\tau,\nu)|^2\, d\nu \right)^{\frac{1}{2}} d\tau$$

$$\leq \left(\int_{-\infty}^{\infty} \int_{-\infty}^{\infty} t^2 |\lambda(t,\tau)|^2\, dt\, d\tau \int_{-\infty}^{\infty} \int_{-\infty}^{\infty} \nu^2 |\Phi(\tau,\nu)|^2\, d\nu\, d\tau \right)^{\frac{1}{2}}. \tag{E.2}$$

By Parseval's thereom,

$$\int\limits_{-\infty}^{\infty} |\lambda(t,\tau)|^2\, d\tau = \frac{1}{2\pi}\int\limits_{-\infty}^{\infty} |\phi(t,\omega)|^2\, d\omega \qquad (E.3a)$$

and

$$\int\limits_{-\infty}^{\infty} |\lambda(t,\tau)|^2\, dt = \frac{1}{2\pi}\int\limits_{-\infty}^{\infty} |\Phi(\tau,\nu)|^2\, d\omega. \qquad (E.3b)$$

Substituting Eq. E.3 into Eq. E.2 yields

$$\frac{1}{2}\int\limits_{-\infty}^{\infty}\int\limits_{-\infty}^{\infty} |\Phi(\tau,\nu)|^2\, d\nu\, d\tau$$

$$\leq \left(\int\limits_{-\infty}^{\infty}\int\limits_{-\infty}^{\infty} t^2|\phi(t,\omega)|^2\, dt\, d\omega \int\limits_{-\infty}^{\infty}\int\limits_{-\infty}^{\infty} \nu^2|\Phi(\tau,\nu)|^2\, d\nu\, d\tau\right)^{\frac{1}{2}}. \; (E.4)$$

Since $\int\int |\phi(t,\omega)|^2\, dt\, d\omega = \int\int |\Phi(\tau,\nu)|^2\, d\tau\, d\nu$, we have $\frac{1}{2} \leq \sigma_t \Sigma_\nu$.
By similar reasoning, $\frac{1}{2} \leq \sigma_\omega \Sigma_\tau$.

Direct computation of the variances shows that if $\phi(t,\omega)$ is a 2-D gaussian (Eq. 2.6.7), then these inequalities are satisfied with equality. Showing the converse is somewhat more involved. If these inequalities are satisfied with equality, then from the classical uncertainty relation and the proof above, it follows that $\Phi(\tau,\nu)$ is Gaussian in each of its variables. In other words,

$$\Phi(\tau,\nu) = e^{-[a(\nu)\tau^2 + b(\nu)]}$$
$$= e^{-[c(\tau)\nu^2 + d(\tau)]}, \qquad (E.5)$$

for all τ and ν, where $a > 0$ and $c > 0$. Thus, $a(\nu)\tau^2 + b(\nu) = c(\tau)\nu^2 + d(\tau)$. Setting $\nu = 0$ and $\tau = 0$ shows that $b(\nu) = c(0)\nu^2 + d(0)$ and $d(\tau) = a(0)\tau^2 + b(0)$, respectively, so

$$a(\nu)\tau^2 + c(0)\nu^2 + d(0) = c(\tau)\nu^2 + a(0)\tau^2 + b(0). \qquad (E.6)$$

Twice differentiating this w.r.t. τ and ν gives $a''(\nu) = c''(\tau)$ for all τ and ν; thus they are constant. Taylor expanding $a(\nu)$ and $c(\tau)$, substituting into Eq. E.6, and equating terms shows that

$$\Phi(\tau,\nu) = e^{-[a(0)\tau^2 + c(0)\nu^2 + \frac{1}{2}a''(0)\tau^2\nu^2 + b(0)]} \qquad (E.7)$$

By the symmetry of the two domains, $\phi(t,\omega)$ must have the same form. Together, these imply that

$$\lambda(t,\tau) = e^{-[\alpha_1 t^2 + \beta_1 \tau^2 + \gamma_1 t^2/\tau^2 + \delta_1]}$$
$$= e^{-[\alpha_2 t^2 + \beta_2 \tau^2 + \gamma_2 \tau^2/t^2 + \delta_2]}, \qquad (E.8)$$

for all t and τ. Taking the logarithm of Eq. E.8, clearing of fractions, and equating terms shows that $\gamma_1 = \gamma_2 = 0$. Thus, $a''(0) = 0$ in Eq. E.7, which implies Eq. 2.6.7, as desired. ///

2.7. Satisfying the design criteria

From the last theorem, we see that a two-dimensional gaussian transform kernel gives the best time-frequency locality for a given smoothness. The resulting representation will be called the *Gaussian transform* of the signal. † By specifying σ_T^2 $(= 2\sigma_t^2)$ and σ_Ω^2 $(= 2\sigma_\omega^2)$ for this kernel we are, in effect, selecting a particular time and frequency scale for the transform. We may choose any values we wish provided $\sigma_T\sigma_\Omega \geq \frac{1}{2}$ (positivity), and the resulting transform will best satisfy all our design properties. The result is clearly a generalization of the solution in the stationary case, where a gaussian convolution kernel of different sizes selected different spectral scales.

---

† We have chosen this name for obvious reasons. This risks, however, confusion with the *Gauss-Weierstrass transformation* [see. Hille 1948]. In fact, the Gaussian transform of the signal $x(t)$ is the two-dimensional Gauss-Weierstrass transformation of the Wigner distribution $W_x(t)$ [see De Bruijn 1967].

When $\sigma_T \sigma_\Omega = \frac{1}{2}$, this transform is equivalent to a spectrogram using a gaussian window. For larger values of $\sigma_T \sigma_\Omega$, this transform is equivalent to convolving such a spectrogram with a 2-D gaussian.

As a note on its implementation, this last fact was used to compute the figures below. A more direct method would be to compute the Wigner distribution and then perform the 2-D convolution specified in Eq. 2.5.2. This is not very efficient in a digital implementation, however, since the Wigner distribution has to be computed at high sampling rates to avoid aliasing. *

By performing a convolution on a spectrogram, far fewer time and frequency samples need to be computed, since the spectrogram is already a smoothed version of the Wigner distribution. Further, since the gaussian kernel is uncorrelated in time and frequency, the 2-D convolution is separable, and can be performed as separate 1-D convolutions in the time and frequency directions, resulting in a relatively inexpensive computation.

2.8. Directional time-frequency transforms

So far, we have assumed that the time-frequency energy representation was non-directional in the sense that the covariances $\sigma_{t\omega}$ and $\Sigma_{\tau\nu}$ of the transform kernel were both zero. We shall now examine the consequences of lifting this condition. We begin with an example. Consider the two transforms specified by the kernels

$$\phi_1(t,\omega) = e^{-(t^2 + 2t\omega + \omega^2)}$$

and

$$\phi_2(t,\omega) = e^{-(t^2 - 2t\omega + \omega^2)}.$$

---

* In general, the Wigner distribution must be sampled in time at twice the Nyquist rate of the signal [Claasen & Mecklenbräuker 1980b].

These transforms have identical σ_t and σ_ω, but differ in the sign of $\sigma_{t\omega}$. Figure 2.10 shows their concentration ellipses, and Figure 2.11 gives the transform of the chirp $e^{i\frac{1}{2}t^2}$ for these two cases. Notice that the second transform broadens the chirp much more than the first, which should be evident from the concentration ellipses. The opposite would be true for the chirp $e^{-i\frac{1}{2}t^2}$. These transforms are directionally sensitive, and using σ_t and σ_ω as the sole measures of time-frequency resolution is obviously inadequate in such cases.

Why consider transforms with such behavior? One answer is to provide a general treatment of time-frequency locality. Another answer is that it is evidently possible to obtain better time-frequency resolution for some signals if the transform is directionally 'tuned' to them than otherwise. This would mean that, in general, we would need a family of transforms each tuned to a preferred time-frequency orientation.

The theory of directional transforms is greatly simplified by a rotation of co-ordinates. Let

$$R_\theta(t,\omega) = \begin{pmatrix} \cos\theta & \sin\theta \\ -\sin\theta & \cos\theta \end{pmatrix} \begin{pmatrix} t \\ \omega \end{pmatrix} \tag{2.8.1}$$

be the operator that rotates a point θ radians in the time-frequency plane. Given a time-frequency representation $F_x(t,\omega)$ of a signal $x(t)$, we can consider the rotated representation formed by the composition $F_x R_\theta(t,\omega)$. Is this the time-frequency representation of an actual signal? The answer is yes; if

$$x_\theta(t) = \frac{1}{2\pi\sqrt{\cos\theta}} e^{\frac{it^2\tan\theta}{2}} \int_{-\infty}^{\infty} X(\omega)e^{i\left(\frac{\omega^2\tan\theta}{2} + \frac{\omega t}{\cos\theta}\right)} d\omega, \tag{2.8.2}$$

then $W_{x_\theta} = W_x R_\theta$ [see Van Trees 1971]. So if F_x has the kernel $\phi(t,\omega)$ and if G_x has the kernel $\phi(t,\omega)R_\theta$, then $G_{x_\theta} = F_r R_\theta$. In other

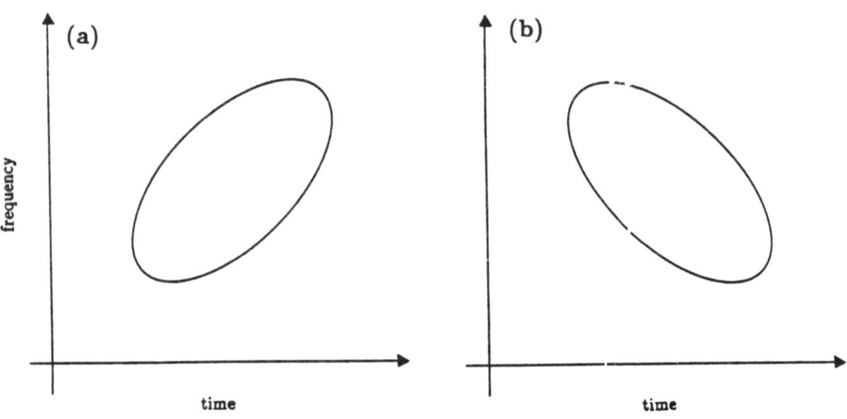

Figure 2.10. *Concentration ellipses for transform kernels with complementary orientation selectivity. (a) Concentration ellipse for* $\phi_1(t,\omega) = e^{-(t^2+2t\omega+\omega^2)}$. *(b) Concentration ellipse for* $\phi_2(t,\omega) = e^{-(t^2-2t\omega+\omega^2)}$.

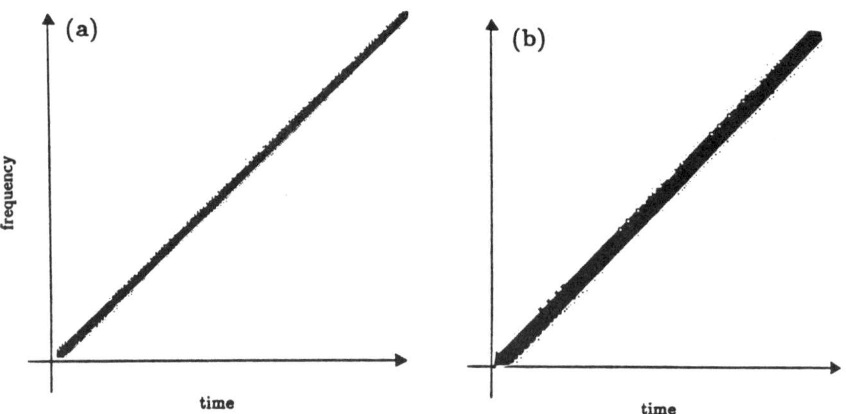

Figure 2.11. *Directional transforms for a linear chirp* $e^{i\frac{1}{2}t^2}$. *(a) Transform has kernel in Figure 2.10a. (b) Transform has kernel in Figure 2.10b. The second transform broadens this chirp much more than the first, which should be evident from their concentration ellipses.*

words, Eq. 2.8.2 rotates the signal by θ radians in time-frequency, thus the transform with the rotated kernel applied to this signal will give the desired effect.

Relative to these new co-ordinates we can generalize some of the measures of the previous sections. For example, consider

$$\pi_\theta(t) = \frac{1}{2\pi} \int\limits_{-\infty}^{\infty} F_x R_\theta(t,\omega)\, d\omega. \tag{2.8.3}$$

This is the marginal of the rotated transform along ω. It follows that the time and frequency marginals (Eq. 2.4.2) of $F_x(t,\omega)$ satisfy $\pi_1 = \pi_{\theta=0}$ and $\pi_2 = 2\pi\, \pi_{\theta=\pi/2}$.

If $\pi_\theta(t) = |x_\theta(t)|^2$, then we will say that the transform preserves the marginal relative to the direction θ in time-frequency. Interestingly, the Wigner distribution uniquely meets this requirement for all θ. The proof is a simple generalization of Cohen's result. He showed that a shift-invariant quadratic transform perserves the time marginal, i.e., $\pi_1(t) = |x(t)|^2$, iff $\Phi(\tau,0) = 1$ for all τ. Using $\mathcal{F}[\phi R_\theta] = \Phi R_\theta$, which is easily verified, it follows that $\pi_\theta(t) = |x_\theta(t)|^2$ iff $\Phi R_\theta(\tau,0) = 1$ for all τ. This implies that $\Phi(\tau,\nu) = 1$, which corresponds to the Wigner distribution by Eq. 2.5.3. This is the reason for considering the Wigner distribution 'perfectly localized' and $\phi(t,\omega)$ the 'point spread function' in time-frequency.

The amount of spread in time-frequency direction θ can be measured by the variance

$$\sigma_\theta^2 = \frac{\displaystyle\int\limits_{-\infty}^{\infty}\int\limits_{-\infty}^{\infty} t^2 |\phi R_\theta^{-1}(t,\omega)|^2\, dt\, d\omega}{\displaystyle\int\limits_{-\infty}^{\infty}\int\limits_{-\infty}^{\infty} |\phi R_\theta^{-1}(t,\omega)|^2\, dt\, d\omega}. \tag{2.8.4}$$

In the notation of the previous sections, $\sigma_t = \sigma_{\theta=0}$, $\sigma_\omega = \sigma_{\theta=\pi/2}$, and

$$\sigma_\theta^2 = (\, \cos\theta \quad \sin\theta \,) \begin{pmatrix} \sigma_t^2 & \sigma_{t\omega} \\ \sigma_{t\omega} & \sigma_\omega^2 \end{pmatrix} \begin{pmatrix} \cos\theta \\ \sin\theta \end{pmatrix}. \qquad (2.8.5)$$

Let σ_1^2 be the maximum value and σ_2^2 be the minimum value σ_θ^2. which corresponds to the eigenvalues of the covariance matrix in Eq. 2.8.5. Further, let θ^* be the maximum direction, which corresponds to the eigenvector $\begin{pmatrix} \cos\theta^* \\ \sin\theta^* \end{pmatrix}$ of the eigenvalue σ_1^2. In other words, σ_1 and σ_2 are the maximum and minimum dimensions of the concentration ellipse of $\phi(t,\omega)$, and θ^* is angle of the major axis of concentration ellipse relative to the time axis. These three quantities conveniently specify the time-frequency locality of the transform.

In an analogous manner, we can measure the smoothness of the transform in time-frequency direction θ by

$$\Sigma_\theta^2 = \frac{\displaystyle\int_{-\infty}^{\infty}\int_{-\infty}^{\infty} \tau^2 |\Phi R_\theta^{-1}(\tau,\nu)|^2\, d\tau\, d\nu}{\displaystyle\int_{-\infty}^{\infty}\int_{-\infty}^{\infty} |\Phi R_\theta^{-1}(\tau,\nu)|^2\, d\tau\, d\nu}. \qquad (2.8.6)$$

In the notation of the previous sections, $\Sigma_\tau = \Sigma_{\theta=0}$, $\Sigma_\nu = \Sigma_{\theta=\pi/2}$, and

$$\Sigma_\theta^2 = (\, \cos\theta \quad \sin\theta \,) \begin{pmatrix} \Sigma_\tau^2 & \Sigma_{\tau\nu} \\ \Sigma_{\tau\nu} & \Sigma_\nu^2 \end{pmatrix} \begin{pmatrix} \cos\theta \\ \sin\theta \end{pmatrix}. \qquad (2.8.7)$$

Let Σ_1^2 be the maximum value and Σ_2^2 be the minimum value of Σ_θ^2, and let θ^{**} be the maximum direction. These three quantities conveniently specify the time-frequency smoothness of the transform.

We are now in a position to generalize Theorem E.

Theorem F. *If $F_x(t,\omega)$ is shift-invariant, then $\sigma_1\Sigma_2 \geq \frac{1}{2}$ and $\sigma_2\Sigma_1 \geq \frac{1}{2}$, with equality in both these relations iff*

$$\phi R_{\theta_*}^{-1}(t,\omega) \propto e^{-(t/2\sigma_1)^2} e^{-(\omega/2\sigma_2)^2}. \qquad (2.8.8)$$

Proof: Applying Theorem E to the transform with kernel $\phi R_{\theta_*}^{-1}$, we have $\frac{1}{2} \leq \sigma_2\Sigma_\tau \leq \sigma_2\Sigma_1$. Similarly, with the kernel $\phi R_{\theta_{**}}$, $\frac{1}{2} \leq \sigma_t\Sigma_2 \leq \sigma_1\Sigma_2$. The righthand inequalities are satisfied with equality iff $\theta^* = \theta^{**}$. It follows from Theorem E that Eq. 2.8.8 is a necessary and sufficient condition that all these inequalites are satisfied with equality. ///

Generalizing Theorem D requires that we use the directional variance of $\phi(t,\omega)$ not $|\phi(t,\omega)|^2$, i.e.,

$$\frac{\displaystyle\int_{-\infty}^{\infty}\int_{-\infty}^{\infty} t^2 \phi R_\theta^{-1}(t,\omega)\, dt\, d\omega}{\displaystyle\int_{-\infty}^{\infty}\int_{-\infty}^{\infty} \phi R_\theta^{-1}(t,\omega)\, dt\, d\omega}. \qquad (2.8.9)$$

We define σ_I^2 and σ_{II}^2 as the maximum and minimum values of this variance, and Θ^* as the maximum direction.

Theorem G. *Let $F_x(t,\omega)$ be positive and shift-invariant. Then $\sigma_I\sigma_{II} \geq \frac{1}{2}$.*

Proof: Apply Theorem D to the signal $x_{-\Theta^*}(t)$ and the transform with kernel $\phi R_{\Theta^*}^{-1}$. ///

Corollary. *If $F_x(t,\omega)$ is positive and shift-invariant, then*

$$\begin{vmatrix} \sigma_t^2 & \sigma_{t\omega} \\ \sigma_{t\omega} & \sigma_\omega^2 \end{vmatrix} \geq \frac{1}{2}.$$

From Theorem F, we see that a two-dimensional gaussian transform kernel gives the best time-frequency locality for a given smoothness. In this general case, however, the gaussian kernel may be correlated in time and frequency, i.e. its concentration ellipse may be oriented obliquely in the time-frequency plane. By specifying σ_I^2 ($= 2\sigma_1^2$), σ_{II}^2 ($= 2\sigma_2^2$), and θ^* for this kernel we are, in effect, selecting a particular time-frequency scale for the transform. By Theorem G, we may choose any values we wish provided $\sigma_I \sigma_{II} \geq \frac{1}{2}$, and the resulting transform will best satisfy all our design properties.

When $\sigma_I \sigma_{II} = \frac{1}{2}$, this transform is equivalent to a spectrogram with a rotated gaussian window $g_{\theta^*}(t)$ [cf. Riley 1983, Dungeon 1984]. For larger values of $\sigma_I \sigma_{II}$, this transform is equivalent to convolving such a spectrogram with a 2-D gaussian.

2.9. A speech example

In this section we examine a particular utterance, comparing the various signal representations discussed above. The utterance is /wioi/ taken from "We owe Eve a dollar", as produced by an adult male. This utterance has some rapid F2 motion, which makes it useful as an example of non-stationary behavior in speech.

Figure 2.12a,b show the traditional wideband and narrowband spectrograms for this utterance. These are spectrograms computed with gaussian windows of standard deviation 1 msec and 15 msecs, respectively. The wideband spectrogram shows vertical striations spaced at the pitch period. The narrowband spectrogram shows horizontal striations spaced at the fundamental frequency. They are both due to the voiced excitation. Figure 2.12c shows a spectrogram whose win-

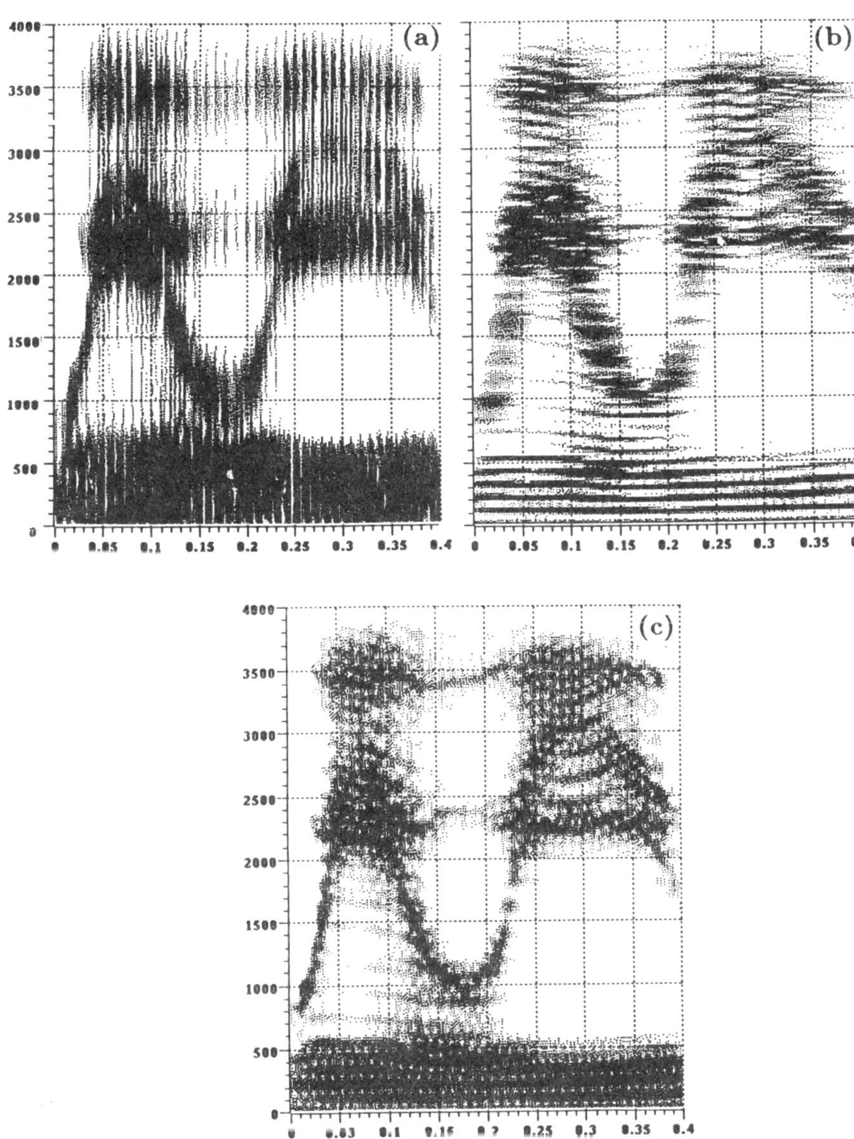

Figure 2.12. Log magnitude spectrograms of the utterance /wioi/. (a) Wideband (gaussian window standard deviation of 1 msec). (b) Narrowband (standard deviation of 15 msec). (c) Intermediate band (standard deviation of 4 msec).

dow duration is 4 msec, which is intermediate between the previous two. This window size is matched to the excitation in the following sense. The 2-D gaussian kernel (Eq. 2.6.7) that corresponds to this spectrogram has standard deviations of 2 msec by 20 Hz. These are in the same ratio as 10 msec and 100 Hz, the pitch period and the fundmental frequency, respectively. This choice gives rise to rows and columns of sharp peaks and valleys spaced at the pitch period and the fundamental frequency. We will see in the next chapter why the excitation produces this particular structure.

Figure 2.13 shows the Wigner distribution for this utterance. Compared to Figure 2.12 it looks almost as if the vertical scale has changed, but it has not. This representation is dominated by cross-terms that give 'echoes' of the formants in initially suprising places. But remember that the sum of two complex exponetials at different frequencies gave rise to a cross-term half-way between them that had greater amplitude than the original terms (Figure 2.8). Evidently, the Wigner distribution itself gives a confusing picture of multi-component signals such as speech.

Figure 2.14 shows the time-frequency autocorrelation function, the 2-D fourier transform of the Wigner distribution, for this utterance in the neighborhood of the origin. Notice the repeated pattern in rows and columns spaced at the pitch period and the fundmental frequency. In Chapter 3 we will see that this pattern can be exploited in understanding how to suppress the excitation.

Figure 2.15 shows the Gaussian transform of this signal using a kernel of a scale chosen to suppress the excitation. The pitch striations are removed, leaving smooth time-frequency ridges that correspond to the formants. The ridges are quite sharp, although it is somewhat difficult to appreciate this in the half-toned picture, Figure 2.15a. The 3-D plot in Figure 2.15b gives a different perspective on this

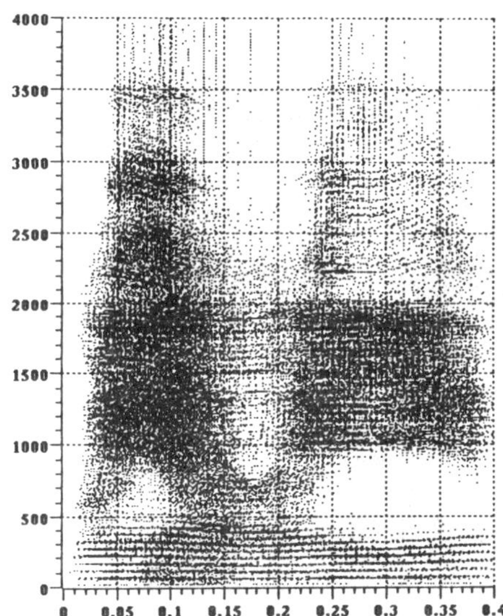

Figure 2.13. *Log magnitude of Wigner distribution. (This is implemented as a pseudo-Wigner distribution using a gaussian window of
standard deviation 40 msec [see Claasen & Mecklenbräuker 1980b].)*

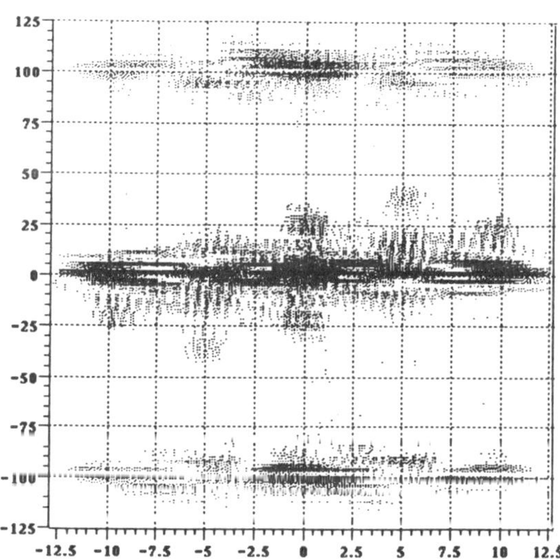

Figure 2.14. *Log magnitude of time-frequency autocorrelation
function in the vicinity of the origin.*

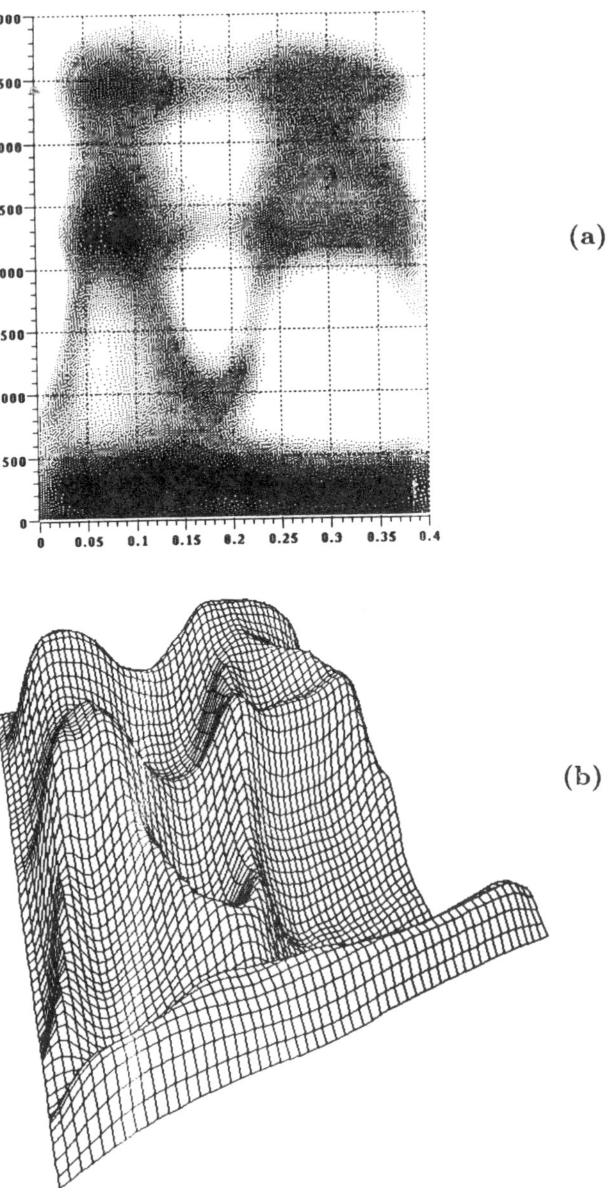

(a)

(b)

Figure 2.15. *Gaussian transform with kernel scales chosen to suppress the excitation, $\sigma_t = 10$ msec and $\sigma_\omega = 100$ Hz. (a) 2-D plot. (b) 3-D plot.*

surface. It shows F1 and parts of F2 quite nicely, although most everything above 2 kHz is considerably distorted in this presentation.

Finally, Figure 2.16 shows directional transforms of this utterance using oriented Gaussian kernels matched to different aspects of the signal. In Figure 2.16a, the kernel orientation is matched to the rising F2. In Figure 2.16b, the kernel orientation is matched to the falling F2. These choices bring out the selected formant peak with high resolution.

In this chapter, we have found that a particular time-frequency energy representation, the Gaussian transform, best satifies a set of properties deemed desirable. There are several free parameters for this representation (σ_t, σ_ω, and θ^*), which determine the scale and directional selectivity of the transform. Deciding what scales are of interest requires a more specific model of the signal. In the next chapter, we adopt such a model.

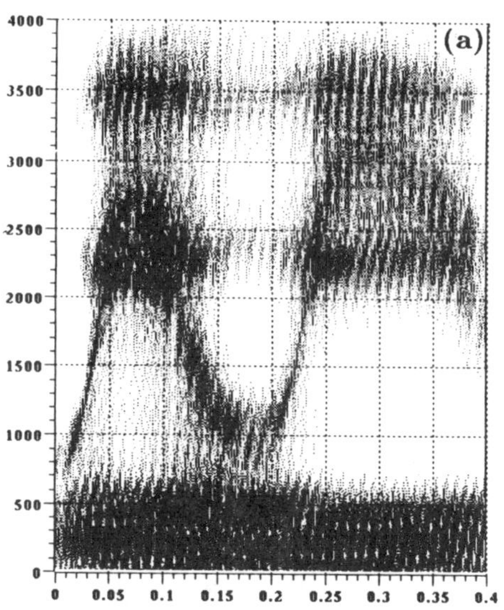

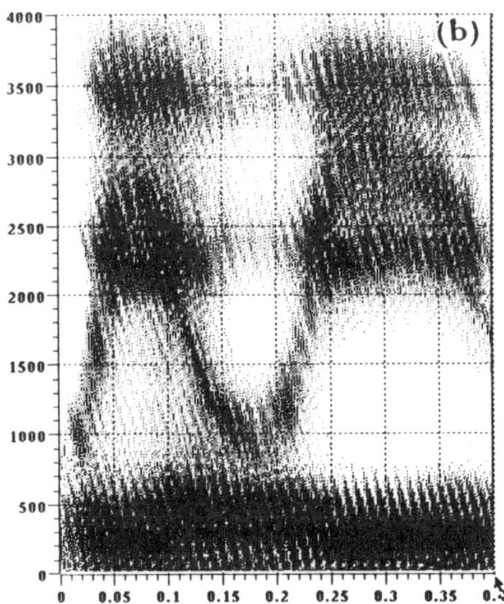

Figure 2.16. *Directional transforms using oriented Gaussian kernels matched to different aspects of the signal. (a) Kernel orientation matched to rising F2. (b) Kernel orientation matched to falling F2.*

3

Time-frequency filtering

In this chapter, we continue the discussion of joint time-frequency energy representations for speech signals. Here we shall make stronger assumptions about the form of the signals. We will introduce a particular model of the time-varying vocal tract, and define its 'transfer function', $H(t,\omega)$. We will show that *time-frequency filtering* can be used to estimate $|H(t,\omega)|^2$, a technique that is essentially a two-dimensional generalization of straight-forward, stationary methods. Further, we will see that $|H(t,\omega)|^2$ is closely related to the time-frequency representations of the previous chapter.

3.1. The stationary case

First, let us re-examine the stationary case. If we adopt a more detailed model of the generation of a stationary speech signal, we can say much more about the cepstral methods discussed in the previous chapter. The linear model [Fant 1960; Flanagan 1972] of vowel production begins by decomposing the speech signal into a vocal source component (e.g. periodic vocal fold vibration) and a vocal tract component, which are treated as independent. The vocal tract is modelled as a linear and quasi-time-invariant filter with excess pressure and volume velocity (of assumed one-dimensional wave motion) being analogous to voltage and current in circuit theory. The distribution of the poles of the filter's system function constitutes the formant description of the vocal tract.

In other words, $H(i\omega)$, the transfer function of the stationary vocal tract, can be approximated by [Flanagan 1972] †

$$H(i\omega) = \sum_{n=1}^{N} \left[z_n H_n(i\omega) + z_n^* H_{-n}(i\omega) \right], \qquad (3.1.1)$$

where $H_n(s)$ consists of a simple pole at $s_n = \alpha_n + i\omega_n$,

$$H_n(i\omega) = \frac{1}{i\omega - (\alpha_n + i\omega_n)}, \qquad (3.1.2)$$

and z_n is the residue at the nth pole,

$$z_n = \frac{\prod_k s_k s_k^*}{2i\omega_n \prod_{k \neq n} \left[(\alpha_k - \alpha_n)^2 + (\omega_k^2 - \omega_n^2) + 2i\omega_n(\alpha_k - \alpha_n) \right]}. \qquad (3.1.3)$$

We associate a formant with each pole, or more precisely, with each pair of poles, since they occur in conjugate pairs, i.e., $s_{-n} = s_n^*$, given the impulse response of the vocal tract is real. The impulse response of the stationary vocal tract, in fact, is

$$h(t) = \sum_{n=1}^{N} \left[z_n h_n(t) + z_n^* h_{-n}(t) \right], \qquad (3.1.4)$$

where

$$h_n(t) = e^{s_n t} u(t). \qquad (3.1.5)$$

---

† This is the parallel formulation. The serial formulation, $H(i\omega) = k \prod_n H_n(i\omega) H_{-n}(i\omega)$ is also often used. The former is the partial fraction expansion of the latter.

In this linear time-invariant model, it follows that the spectrum of the excitation and the vocal tract transfer function combine by multiplication in the power spectrum and addition in the log spectrum. This fact leads to a simple procedure for separating the excitation and the vocal tract transfer function in certain (idealized) cases.

Suppose the the excitation is an impulse train, which is a very simple model of constant pitch, voiced excitation. In this case, the spectrum of the excitation is also an impulse train, and thus, the speech spectrum is a uniformly sampled version of the vocal tract transfer function. If the sampling were unaliased (i.e., the pitch is low enough relative to the highest transfer function quefrencies) the original transfer function can be exactly recovered by ideal low-pass filtering the spectrum, by the sampling theorem [Bracewell 1978]. But this is just cepstral smoothing using, in this very idealized case, a rectangular cepstral window [Oppenheim 1969; Oppenheim & Shafer 1975].

Let us examine this result more closely. The formulation here will be in terms of the power spectrum and its transform, the autocorrelation function, instead of the more usual log spectrum and its transform, the cepstrum, since the former generalizes more easily to the time-varying case. Since the term 'cepstral filtering' is, properly speaking, reserved for filtering operations on the log magnitude spectrum, we shall refer to analogous operations on the power spectrum as *autocorrelation filtering*. The results in the stationary case are similar in either formulation. †

––––––––––––––––––––

† Cepstral and autocorrelation filtering can both be used to separate signal components that arise at different scales in the frequency domain. Cepstral filtering is most appropriate when the signal components combine by convolution in the time domain, autocorrelation filtering when they combine by addition. Both approaches can be used for speech, since we can use either a serial or parallel formulation of the vocal tract model.

If $x(t)$ represents the excitation, $h(t)$ the impulse response of the vocal tract, and $y(t)$ the output speech signal, then in terms of power spectra and transfer function, $|Y(\omega)|^2 = |H(i\omega)|^2 |X(\omega)|^2$, or in terms of autocorrelation functions,

$$A_y(\tau) = \int_{-\infty}^{\infty} A_x(t) A_h(\tau - t)\, dt. \tag{3.1.6}$$

Let the excitation be an impulse train, $I(t; T) = \sum_k \delta(t - kT)$. Then

$$A_I(\tau) = \frac{2\pi}{T} \sum_{k=-\infty}^{\infty} \delta(\tau - kT). \tag{3.1.7}$$

Thus from Eq. 3.1.6, we have

$$A_y(\tau) = \frac{2\pi}{T} \sum_{k} A_h(\tau - kT). \tag{3.1.8}$$

Provided the duration of $A_h(\tau)$ is small enough that the terms in Eq. 3.1.8 do not overlap, $A_h(t)$ and thus $|H(i\omega)|^2$ can be recovered by windowing $A_y(\tau)$ with a rectangular window centered on the origin and of duration T (see Figure 3.1).

Let us examine the form of $A_h(\tau)$. Assume for now that the vocal tract transfer function consists of only a single pole, i.e., its impulse response has the form of Eq. 3.1.5. Then

$$A_{h_n}(\tau) = \int_{-\infty}^{\infty} e^{s_n(\tau+t)} u(\tau + t) e^{s_n^* t} u(t)\, dt$$

$$= e^{s_n \tau} \int_{-\infty}^{\infty} e^{2\alpha_n t} u(\tau + t) u(t)\, dt$$

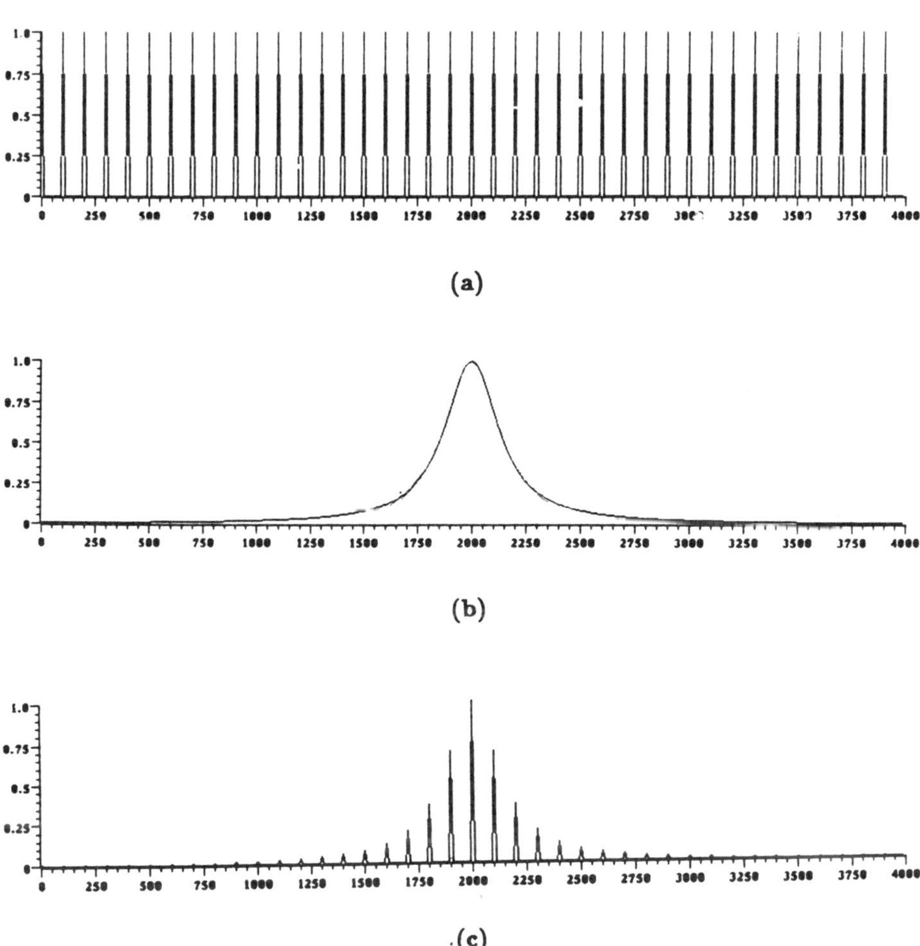

Figure 3.1. Recovering the transfer function by autocorrelation filtering. (a) Spectrum of the excitation modelled as an impulse train (10 msec period). (b) Square magnitude of the transfer function, which in this simple example is a single pole of 300 hz bandwidth. (c) Power spectrum, the product of '(a)' and '(b)'. Cepstral filtering uses the log spectrum instead. The approach here generalizes more easily to the time-varying case. (**continued...**)

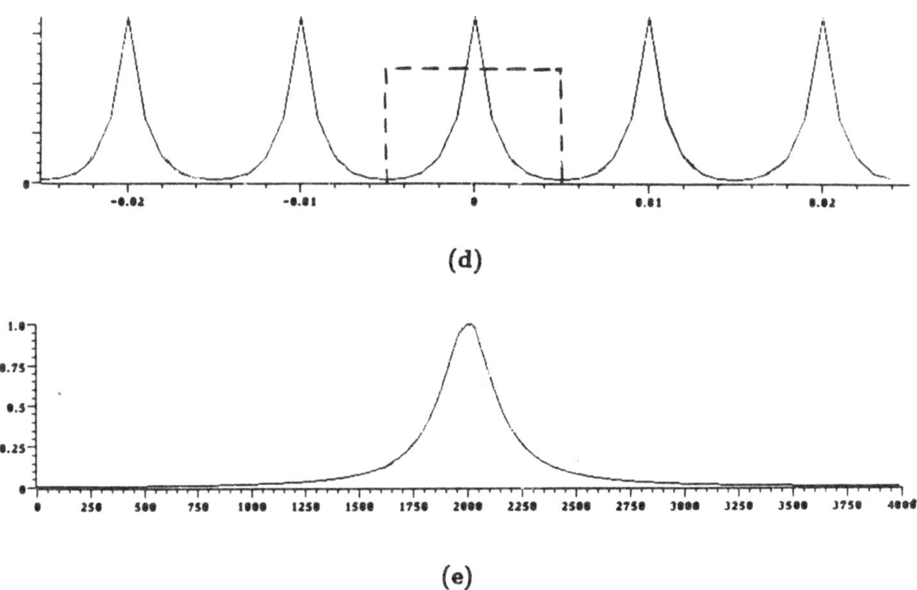

(d)

(e)

Figure 3.1 (continued). *Recovering the transfer function by auto-correlation filtering. (d) Magnitude of the autocorrelation function, the (inverse) fourier transform of '(c)'. Dashed lines show the rectangular window. (e) Fourier transform of the windowed autocorrelation function, which very nearly recovers the transfer function '(b)' in this idealized case (the effect of the slight overlap of the terms in '(d)' is negligible).*

$$= e^{s_n t} \int_{max(-\tau,0)}^{\infty} e^{2\alpha_n t}\, dt$$

$$= \frac{1}{\beta_n} e^{\alpha_n |\tau|} e^{i\omega_n \tau}, \qquad (3.1.9)$$

where $\beta_n = -2\alpha_n$ is the (half-power) bandwidth of the pole. Thus, provided this bandwidth is large enough, the overlap in the terms in Eq. 3.1.8 will be negligible, and windowing $A_y(\tau)$ will very nearly recover $A_h(\tau)$ and hence $|H(i\omega)|^2$. †

The analysis of the multiple pole case follows from superposition. Provided the poles are not closely spaced relative to their bandwidths, ‡

$$|H(i\omega)|^2 \approx \sum_{n=1}^{N} |z_n|^2 \left[|H_n(i\omega)|^2 + |H_{-n}(i\omega)|^2 \right], \qquad (3.1.11)$$

from Eq. 3.1.1 and Eq. 3.1.2, hence

$$A_h(\tau) \approx \sum_{n=1}^{N} \frac{|z_n|^2}{2\beta_n} e^{\alpha_n |\tau|} \cos \omega_n \tau, \qquad (3.1.12)$$

from Eq. 3.1.9. From this equation and Eq. 3.1.8, we see that windowing the autocorrelation function of the output speech signal can still be used to recover the transfer function when the bandwidths are large enough that aliasing is negligible.

A few changes to this model make it more realistic. First, the spectrum of constant voiced excitation is somewhat better modelled as

---

† The phase of the transfer function can be found, if desired, from its magnitude, since this model is minimum phase [see Oppenheim & Shafer 1975].
‡ The analysis in terms of log spectra and cepstra does not require this proviso, since convolutions in the time domain transform (exactly) to sums in the cepstral domain. This is an advantage of the cepstral approach.

an impulse train that drops off at 12DB per octave [Flanagan 1972]. This trend can be removed by spectral pre-emphasis.

Second, the sampling is usually significantly aliased, which is a more serious problem. In this case, we can recover only a low-pass version of the transfer function. A rectangular window is a poor choice in this case, since its transform rings for a considerable duration in the frequency domain. The gaussian is a good choice, because it has minimal bandwidth for a given window duration, as indicated in the previous chapter. (see Figure 3.2). Typically, the standard deviation of the gaussian window is selected about equal to the pitch period.

3.2. Non-stationary vocal tract

Let us now consider the case where the vocal tract configuration is not necessarily static. The goal is to recover the "time-varying transfer function" of the vocal tract from the signal and remove the excitation, as we did in the stationary case.

Unfortunately, there is no widely accepted, satisfactory definition of the transfer function for a time-varying linear filter, although there have been many proposals [e.g., see Lui 1971; Loynes 1968; Page 1952; Saleh & Subotic 1985; Zadeh 1950]. We shall avoid this difficulty by constraining the form of the transfer function; we shall allow non-stationarity, but only in certain well-behaved ways.

The vocal tract, of course, is not an arbitrary time-varying filter; it is constrained by the physical properties of the articulators. Jospa[1982,1984] has investigated the physics of the non-stationary vocal tract analytically, and found that under certain reasonable physical assumptions it is possible to generalize the notion of a formant to the time-varying case. Essentially, he replaces the assumption of a static vocal tract configuration by the assumption that the

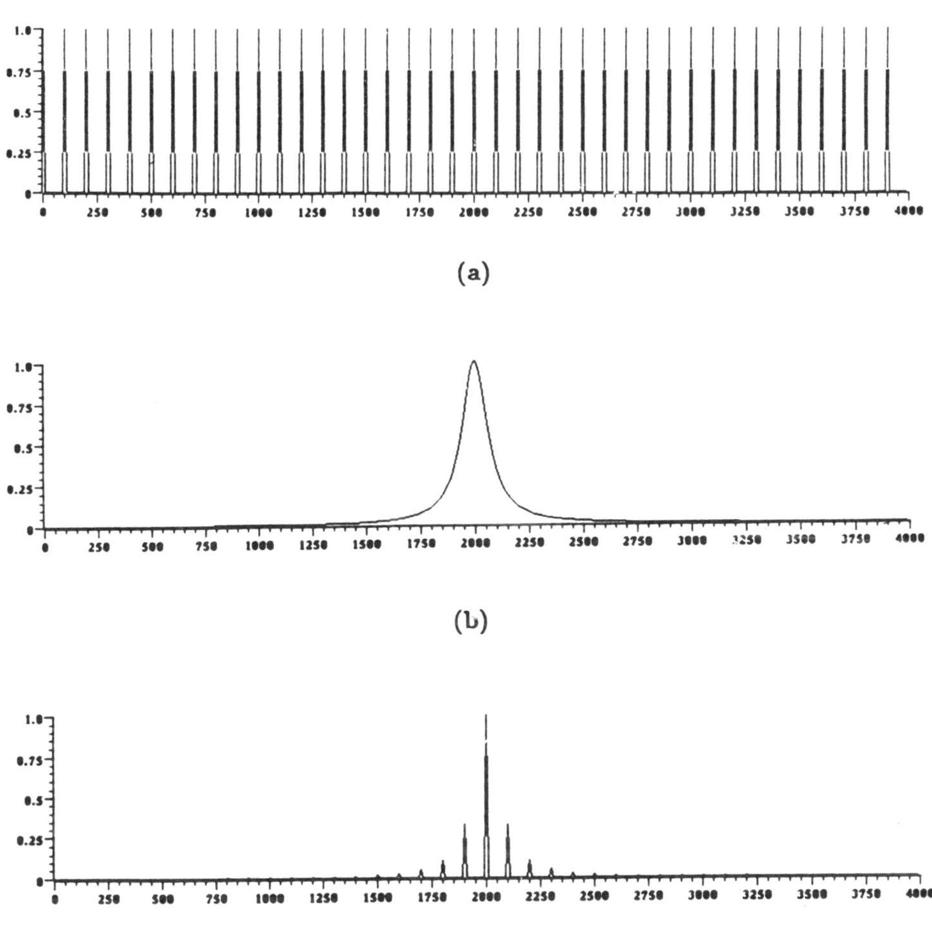

(a)

(b)

(c)

Figure 3.2. *Estimating 'aliased' transfer function. (a) Spectrum of excitation modelled as an impulse train (10 msec period). (b) Square magnitude of the transfer function, a single pole of 150 Hz bandwidth. This has higher 'quefrencies' than the previous example; '(a)' undersamples it in this case. (c) Power spectrum, the product of '(a)' and '(b)'.* (**continued...**)

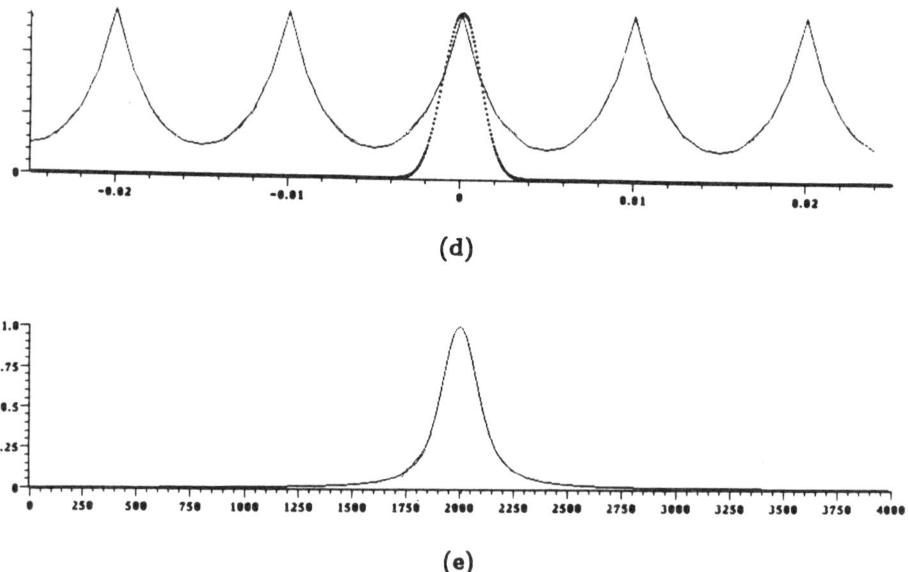

(d)

(e)

Figure 3.2 (continued). *Estimating 'aliased' transfer function. (d) Magnitude of the autocorrelation function, the (inverse) fourier transform of '(c)'. Dotted line show the gaussian window. (e) Fourier transform of the windowed autocorrelation function, which recovers a low-pass version of the transfer function '(b)'.*

deformations are slow enough to satisfy the condition of adiabatic approximation, which he indicates appears to be generally valid from cine X-ray measurements.

We can thus define the impulse response, $h(t, a)$, for a time-varying "resonance" of the vocal tract to an impulse, $\delta(t - a)$, at time a as:

$$h(t, a) = e^{\int_a^t [-2\beta_0 + i(\omega_0 + \gamma(\tau))]\, d\tau} u(t - a), \qquad (3.2.1)$$

where we assume the formant bandwidth β_0 is fixed, and the formant center frequency is ω_0 at $t = 0$. Note that Eq. 3.2.1 reduces to the usual definition of the impulse response of a formant if the time-varying modulation frequency, $\gamma(t)$, is zero.

In Jospa's model, the bandwidth varies somewhat with rate of change of vocal tract area, which we shall treat as negligible. Regarding these bandwidth variations, Fant [1980] believes they "...are of academic rather practical significance. Of greater importance is probably the mere fact that a rapid transition of a formant creates a special perceptual 'chirp' effect."

It will be convenient to examine a more general class of impulse responses than in Eq. 3.2.1. Consider the impulse response

$$h(t, a) = h_0(t - a)e^{i \int_a^t \gamma(\tau)\, d\tau}, \qquad (3.2.2)$$

where $h_0(t)$ is the impulse response of a linear time-invariant (LTI) system and $\gamma(0) = 0$. Eq. 3.2.1 has this form with $h_0(t) = e^{(-2\beta_0 + i\omega_0)t} u(t)$. We call this a *frequency-modulated filter*. We shall study this kind of filter in the next several sections, since it is possible to generalize the notion of a transfer function for it and it is possible to estimate this transfer function by generalizing the autocorrelation filtering methods described above. Of course, an FM filter models only a single pole; we shall take up the multiple pole model of the complete vocal tract transfer function in a later section.

How then can we represent the time-varying transfer function of an FM filter? An intuitively appealing candidate is

$$H(t,\omega) = H_0[i(\omega - \gamma(t))], \qquad (3.2.3)$$

where $H_0(i\omega)$ is the transfer function of the corresponding stationary filter with impulse response $h_0(t)$ (Eq. 3.2.2). In terms of how we might want to visualize the transfer function of an FM filter, this seems attractive; it is just the stationary transfer function shifted at each time by the local modulation frequency $\gamma(t)$. For a time-varying formant pole, $H(t,\omega)$ would have the form of a stationary pole in each frequency cross-section with center frequency $\omega_0 + \gamma(t)$ and fixed bandwidth β_0.

For our purposes, the most important properties that the definition of the time-varying transfer function of a formant should satisfy are practical ones — it should provide phonetically relevant information about the signal, and it should be computable from the signal. The representation in Eq. 3.2.3 satisfies these properties since it is a simple generalization of the stationary case, which is already understood, and it can be estimated from the signal by methods we will describe shortly.

The transfer function of an LTI filter, however, also has some nice theoretical properties that would be desirable when generalized to the time-varying case. In particular, the transfer function $H_0(i\omega)$ of an LTI filter, $y(x) = T_0[x(t)]$: (1) specifies the eigenvalues for the filter's eigenfunctions, i.e.,

$$T_0[e^{i\omega t}] = H_0(i\omega)e^{i\omega t}; \qquad (3.2.4)$$

and (2) is the ratio of the spectrum of the output over the spectrum of the input, i.e.,

$$H_0(i\omega) = \frac{Y(\omega)}{X(\omega)}. \qquad (3.2.5)$$

The first property does generalize to the FM case. Consider the functions

$$\varphi_\omega(t) = e^{i\int_0^t [\omega+\gamma(\tau)]\,d\tau}. \tag{3.2.6}$$

These are the eigenfunctions for an FM filter T, with impulse response defined by Eq. 3.2.2. This follows from

$$
\begin{aligned}
T[\varphi_\omega(t)] &= \int_{-\infty}^{\infty} h(t,a)\varphi_\omega(a)\,da \\
&= \int_{-\infty}^{\infty} h_0(t-a)e^{i\int_a^t \gamma(\tau)\,d\tau}\,e^{i\int_0^a (\omega+\gamma(\tau))\,d\tau}\,da \\
&= e^{i\int_0^t \gamma(\tau)\,d\tau} \int_{-\infty}^{\infty} h_0(t-a)e^{i\omega a}\,da \\
&= e^{i\int_0^t \gamma(\tau)\,d\tau} H_0(i\omega)e^{i\omega t} \\
&= H_0(i\omega)\varphi_\omega(t). \tag{3.2.7}
\end{aligned}
$$

Further, we see from Eq. 3.2.7 that $H_0(i\omega)$ specifies the eigenvalues for the eigenfunctions $\varphi_\omega(t)$. The value of $H_0(i\omega)$, however, depends on the choice of the time origin. More generally,

$$T[\varphi_\omega(t)] = H(0,\omega)\varphi_\omega(t) \tag{3.2.8}$$

is time shift-invariant, where $H(t,\omega)$ is defined by Eq. 3.2.3. †

By comparison, some authors have used

$$\bar{H}(t,\omega) = \int_{-\infty}^{\infty} h(t,a)e^{-i\omega(t-a)}\,da \tag{3.2.9}$$

---

† I.e., suppose $\bar{t} = t - \tau$. Let $\bar{H}(\bar{t},\omega)$ and $\bar{H}_0(i\omega)$ be the time-varying transfer function and the corresponding LTI transfer function, respectively, in the new time co-ordinate. Then, $\bar{H}(\bar{t},\omega) = H(\bar{t}+\tau,\omega)$ and $\bar{H}_0(i\omega) = H_0[i(\omega - \gamma(\tau))] = H(\tau,\omega) = \bar{H}(0,\omega)$.

as their definition of the time-varying transfer function [e.g., Zadeh 1950]. The filter's response to a complex exponential $e^{i\omega t}$ is $\hat{H}(t,\omega)e^{i\omega t}$. However, $e^{i\omega t}$ is not, in general, an eigenfunction of a time-varying system, consequently $\hat{H}(t,\omega)$ has limited use.

Saleh & Subotic [1985] have explored generalizing the second property (Eq. 3.2.5) to the time-varying case. They suggest using

$$\tilde{H}(t,\omega) = \frac{F_y(t,\omega)}{F_x(t,\omega)} \tag{3.2.10}$$

as the definition of the time-varying transfer function where $F_x(t,\omega)$ and $F_y(t,\omega)$ are joint time-frequency representations of the input and output signals, respectively. The difficulty with their approach is that the ratio in Eq. 3.2.10, in general, will have different values for different inputs $x(t)$ for a given filter, unlike the LTI case (Eq. 3.2.5). This second property evidently does not generalize well to the time-varying case.

3.3. Time-frequency filtering

The remainder of this chapter is used to show that *time-frequency filtering* can be used to estimate the transfer function of FM filters and, more generally, of the time-varying vocal tract. Time-frequency filtering consists of multiplying the time-frequency autocorrelation function $A_x(\tau,\nu)$ (Eq. 2.5.5) of the signal $x(t)$ with a 2-D window $\Phi(\tau,\nu)$. The 2-D inverse fourier transform of this windowed function,

$$\mathcal{F}^{-1}\left[\Phi(\tau,\nu)A_x(\tau,\nu)\right], \tag{3.3.1}$$

becomes the filtered time-frequency representation. The shape of the window, of course, determines what energy is kept and what is removed in the filtered representation [cf. Flandrin 1984].

This technique is in many ways the time-varying generalization of the autocorrelation filtering methods presented in Section 3.1. The time-frequency autocorrelation takes the place of the autocorrelation function, a 2-D window the place of a 1-D window, and a 2-D inverse fourier transform of a 1-D fourier transform in this generalization.

The representation in Eq. 3.3.1 also specifies a general member of the quadratic transforms presented in the previous chapter, indicating that the two chapters are related. In this chapter, our goal is to show that a member of this class can give a good estimate of the time-varying "transfer function" of the vocal tract. Happily, it turns out that the form of time-frequency window $\Phi(\tau, \nu)$ that gives a good estimate is a 2-D gaussian, which is the same as Eq. 2.6.7. In other words, we end up with the same kind of time-frequency representation as in the previous chapter, which was based there on weaker, but more general goals.

The results of this chapter, then, reinforce and reinterpret those of the previous chapter. Further, the analysis here suggests which scales to choose, decisions that were free parameters of Chapter 2. In particular, for voiced speech, σ_t is matched to the pitch period, and σ_ω is matched to the fundamental frequency.

We have just given the basic result of this chapter. It remains to demonstrate its validity, i.e., that this kind of filtering will give a good estimate of the time-varying vocal tract "transfer function". This requires several steps in which we gradually generalize the form of the filter that models the vocal tract. In Section 3.4, we re-examine the stationary case, this time in terms of the time-frequency auto-correlation function. In Section 3.5, we consider FM filters that have a linearly varying modulation frequency. In Section 3.6, we use a locality argument to generalize these results for quasi-stationary filters and for FM filters that have a smoothly varying modulation frequency, respectively. In Section 3.7, we use a superposition argument to treat the multiple pole case.

3.4. The stationary case — re-examined

So let us assume for now we want to estimate the transfer function of a filter that is time-invariant. We will show how the time-frequency autocorrelation function can be used to produce this estimate.

This will really just be recapitulation of the stationary argument presented in Section 3.1. In fact, $A_h(\tau,0) = A_h(\tau)$, so we see the correspondence is very close. But with the time-frequency autocorrelation function we will be in a position to generalize these results to the time-varying case, so it is worth the effort.

Letting $x(t)$ represent the filter input, $h(t)$ the filter's impulse response, and $y(t)$ the output, we have

$$A_y(\tau,\nu) = \int\limits_{-\infty}^{\infty} A_x(t,\nu)A_h(\tau - t,\nu)\, dt. \qquad (3.4.1)$$

In other words, the time-frequency autocorrelation function $A_y(\tau,\nu)$ consists of the convolution of $A_x(\tau,\nu)$ and $A_h(\tau,\nu)$ along the τ dimension. This is analogous to Eq. 3.1.6.

Let the filter input be an impulse train $I(t;T) = \sum_n \delta(t - nT)$. Then

$$A_I(\tau,\nu) = \int\limits_{-\infty}^{\infty} e^{-i\nu t} \sum_n \delta(t - nT + \tau/2) \sum_m \delta(t - mT - \tau/2)\, dt.$$

Substituting $t' = t - \frac{1}{2}(m + n)T$ and $\tau' = \tau + (m - n)T$,

$$= \sum_n \sum_m \left\{ \int\limits_{-\infty}^{\infty} e^{-i\nu t'} \delta(t' + \tau'/2)\delta(t' - \tau'/2)\, dt' \right\} e^{-i\frac{1}{2}(m+n)T\nu}.$$

The quantity in braces is the time-frequency autocorrelation function of an impulse $\delta(t')$, which is $A_\delta(\tau',\nu) = \delta(\tau')$ [see Classen & Mecklenbräuker 1980a]. Thus,

$$A_I(\tau,\nu) = \sum_n \sum_m \delta(\tau + (m-n)T)e^{-i\frac{1}{2}(m+n)T\nu}.$$

Letting k=n-m,

$$= \sum_n \sum_k \delta(\tau - kT)e^{i\frac{k}{2}T\nu}e^{-inT\nu}$$

$$= \sum_k e^{i\frac{k}{2}T\nu}\delta(\tau - kT)\left\{\sum_n e^{-inT\nu}\right\}.$$

The quantity in braces is the fourier transform of an impulse train $I(t;T)$, which is itself an impulse train $\frac{2\pi}{T}I(\nu;\frac{2\pi}{T})$ [see Bracewell 1978]. Therefore,

$$A_I(\tau,\nu) = \frac{2\pi}{T}\sum_k\sum_n e^{i\frac{k}{2}T\nu}\delta(\tau - kT)\delta(\nu - \frac{2\pi n}{T})$$

$$= \frac{2\pi}{T}\sum_k\sum_n (-1)^{nk}\delta(\tau - kT)\delta(\nu - \frac{2\pi n}{T}). \quad (3.4.2)$$

Eq. 3.4.2 shows that the time-frequency autocorrelation function of an impulse train is a rectangular grid of impulses spaced T apart along τ and $2\pi/T$ apart along ν (see Figure 3.3). † Eq. 3.4.2 is the two-dimensional analog of Eq. 3.1.7.

From Eq. 3.4.1, we have

$$A_y(\tau,\nu) = \frac{2\pi}{T}\sum_k\sum_n (-1)^{nk}A_h(\tau - kT, \frac{2\pi n}{T})\delta(\nu - \frac{2\pi n}{T}), \quad (3.4.3)$$

---

† Siebert [1956] has derived the time-frequency autocorrelation function for a train of pulses of arbitrary shape, a result that is important in the theory of radar. The above result follows formally from this if the pulses are given unit area and approach zero width in the limit.

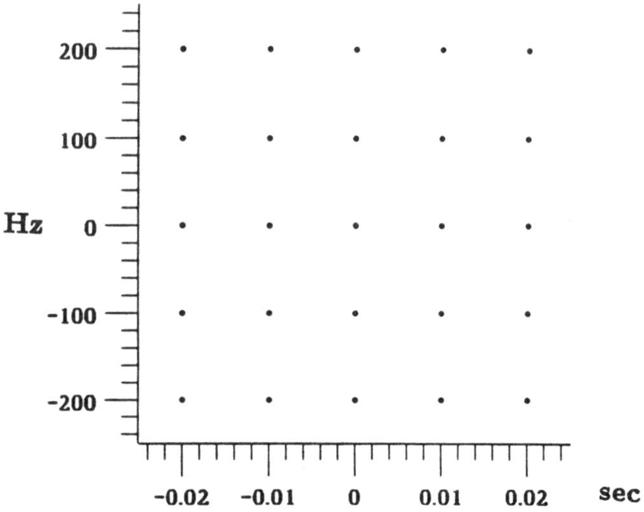

Figure 3.3. *Magnitude of the time-frequency autocorrelation function of an impulse train (10 msec period).*

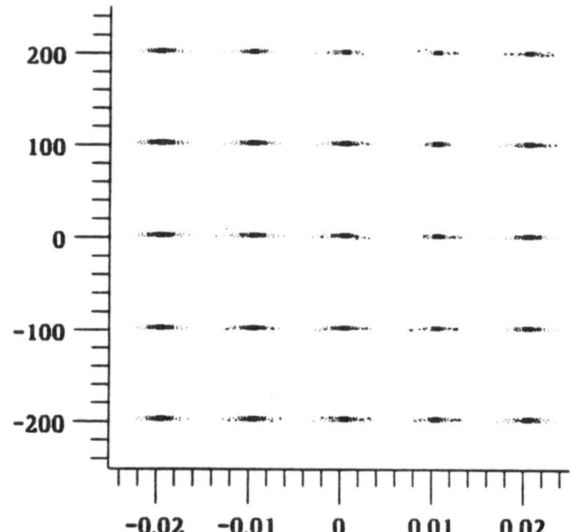

Figure 3.4. *Magnitude of the time-frequency autocorrelation function of the output of an LTI filter excited by an impulse train. In this simple example the filter consists of a single pole of 300 hz bandwidth.*

the two-dimensional analog of Eq. 3.1.8. $A_y(\tau,\nu)$ consists of a rectangular grid of shifted τ slices of $A_h(\tau,\nu)$ (see Figure 3.4).

Provided the terms in Eq. 3.4.3 do not overlap, $A_h(\tau,0)$ can be recovered from $A_y(\tau,\nu)\delta(\nu)$ by windowing it with a rectangular window that is centered on the origin and that has length T, width $2\pi/T$, and height $T/2\pi$ (see Figure 3.5). From $A_h(\tau,0)\delta(\tau)$ we can, in turn, recover $|H(i\omega)|^2$, since

$$
\begin{aligned}
\mathcal{F}^{-\infty}\left[A_h(\tau,0)\delta(\nu)\right] &= \frac{1}{2\pi}\int_{-\infty}^{\infty}\int_{-\infty}^{\infty} A_h(\tau,0)\delta(\nu)e^{i(\nu t-\tau\omega)}\,d\tau\,d\nu \\
&= \int_{-\infty}^{\infty} W_h(t,\omega)\,dt \\
&= |H(i\omega)|^2\,.
\end{aligned}
\tag{3.4.4}
$$

On the other hand, if the terms in Eq. 3.4.3 do overlap somewhat, then a low-pass version of $|H(i\omega)|^2$ can still be recovered, since

$$
\begin{aligned}
\mathcal{F}^{-1}\left[\Phi(\tau,\nu)A_y(\tau,\nu)\right] &\approx \mathcal{F}^{-1}\left[\Phi(\tau,\nu)\frac{2\pi}{T}A_h(\tau,0)\delta(\nu)\right] \\
&= \frac{1}{T}\phi(t,\omega)**|H(i\omega)|^2\,,
\end{aligned}
\tag{3.4.5}
$$

where $\Phi(\tau,\nu)$ is the time-frequency window, and $\phi(\tau,\omega)$ is its two-dimensional inverse fourier transform. In this case, using a rectangular window on the time-frequency autocorrelation function is a poor choice since its transform rings for a considerable duration away from the origin. A gaussian window minimizes this problem.

Let us examine the form of $A_h(\tau,\nu)$ assuming for now that the filter consists of only a single pole, i.e., its impulse response has the form

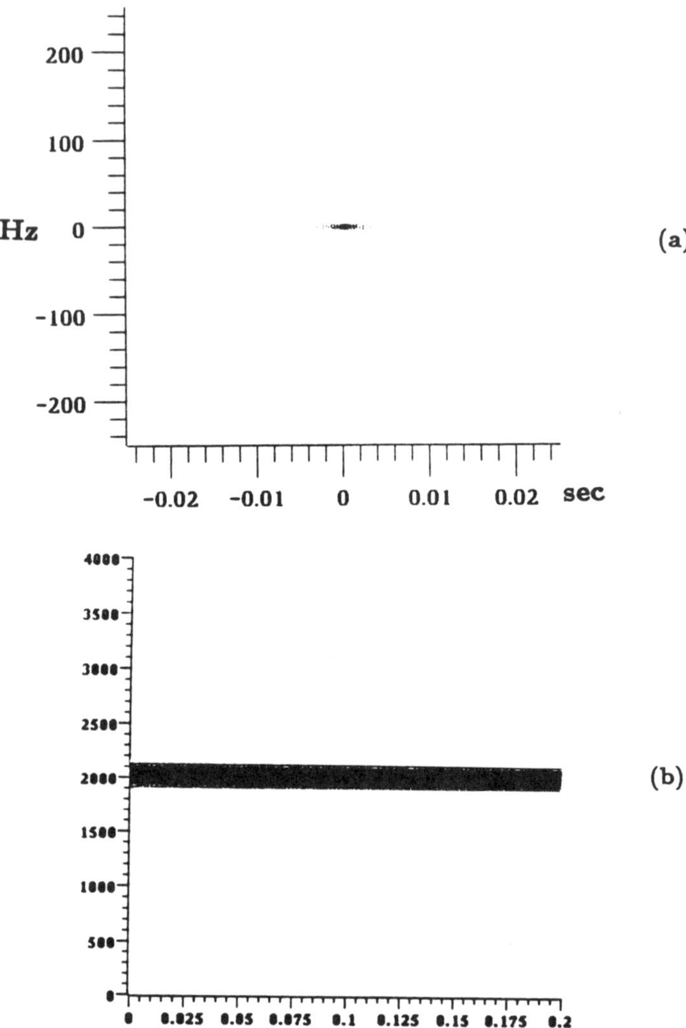

Figure 3.5. *Rectangular window (very nearly) recovers 'unaliased' transfer function. (a) Windowed time-frequency autocorrelation function in Figure 3.4. (b) Square magnitude of transfer function, the 2-D inverse fourier transform of '(a)'. In the 'aliased' case, i.e., if the terms in Figure 3.4 were to overlap significantly, a gaussian window would be more appropriate.*

of Eq. 3.1.5. Then

$$A_{h_n}(\tau,\nu) = \int_{-\infty}^{\infty} e^{s_n(t+\tau/2)} u(t+\tau/2) e^{s_n^*(t-\tau/2)} u(t-\tau/2) e^{-i\nu t}\, dt$$

$$= e^{i\omega_n\tau} \int_{-\infty}^{\infty} e^{2\alpha_n t} u(t-|\tau|/2) e^{-i\nu t}\, dt$$

$$= \frac{e^{(\alpha_n - i\nu/2)|\tau|} e^{i\omega_n\tau}}{\beta_n + i\nu}. \tag{3.4.6}$$

This last equation is the two dimensional analog of Eq. 3.1.9.

Thus, provided the pole bandwidth is large enough, windowing $A_y(\tau,\nu)$ can recover most of $A_h(\tau,\nu)$, and, hence, a low-pass version of $|H(i\omega)|^2$.

3.5. Linearly varying modulation frequency

We now consider the case where we want to estimate the transfer function of an FM filter that has a linearly varying modulation frequency, i.e., $\gamma(t) = mt$ in Eq. 3.2.2. This means

$$h(t,a) = h_0(t-a) e^{i\frac{1}{2}m(t^2-a^2)}. \tag{3.5.1}$$

The previous section was the special case $m = 0$.

Let us find how passing a signal through such a filter modifies its time-frequency autocorrelation function. As usual, we let $x(t)$ represent the input to the filter and $y(t)$ the output. Thus,

$$y(t) = \int_{-\infty}^{\infty} x(a) h(t,a)\, du$$

$$= e^{i\frac{1}{2}mt^2} \int_{-\infty}^{\infty} x(a) e^{-i\frac{1}{2}ma^2} h_0(t-a)\, da. \tag{3.5.2}$$

Letting $\tilde{x}(t) = x(t)e^{-i\frac{1}{2}mt^2}$ and $\tilde{y}(t) = y(t)e^{-i\frac{1}{2}mt^2}$, we have from Eq. 3.5.2 and Eq. 3.4.1,

$$A_{\tilde{y}}(\tau,\nu) = \int_{-\infty}^{\infty} A_{\tilde{x}}(t,\nu)A_{h_0}(\tau - t,\nu)\, dt. \qquad (3.5.3)$$

In other words, the time-frequency autocorrelation function of $\tilde{y}(t)$ consists of the convolution of the time-frequency autocorrelation of $\tilde{x}(t)$ and $h_0(t)$ along the τ dimension.

We are more directly interested in A_x and A_y, than $A_{\tilde{x}}$ and $A_{\tilde{y}}$. But this last transformation in simple, since the time-frequency autocorrelation function has the following nice property: if $\tilde{x}(t) = x(t)e^{-i\frac{1}{2}mt^2}$, then [Van Trees 1971]

$$A_{\tilde{x}}(\tau,\nu) = A_x(\tau,\nu + m\tau). \qquad (3.5.4)$$

In other words, multiplying a signal by a linear chirp *shears* its time-frequency autocorrelation function along the ν dimension (see Figure 3.6).

Combining Eq. 3.5.3 and Eq. 3.5.4, we see that

$$A_y(\tau,\nu) = \int_{-\infty}^{\infty} A_x(t,\nu + m(t - \tau))A_{h_0}(\tau - t,\nu - m\tau)\, dt. \quad (3.5.5)$$

In words, the time-frequency autocorrelation function of a signal passed through the filter in Eq. 3.5.1 can be found by first shearing its input time-frequency autocorrelation function, convolving that with the time-frequency autocorrelation function of $h_0(t)$, and then shearing the output time-frequency autocorrelation function in the opposite direction, all with respect to the ν dimension (see Figure 3.7).

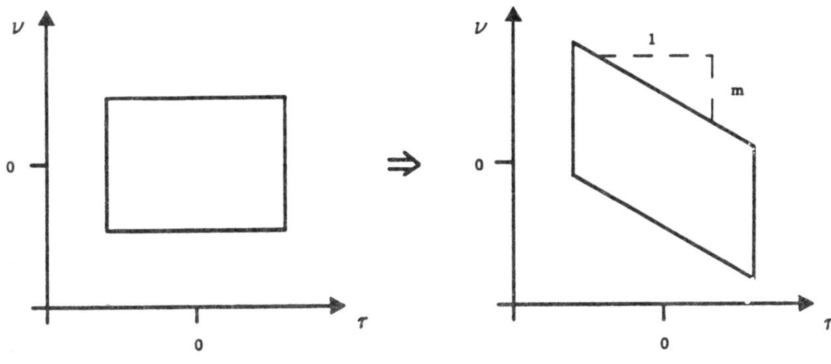

Figure 3.6. *Multiplying a signal $x(t)$ by e^{-imt} shears its time-frequency autocorrelation function: $A_x(\tau, \nu + m\tau)$.*

194z When the filter input is an impulse train $I(t; T)$, the filter output is

$$A_y(\tau, \nu) = \frac{2\pi}{T} \sum_k \sum_n (-1)^{nk} A_{h_0}(\tau - kT, \frac{2\pi n}{T} - mkT)$$

$$\times \delta(\nu - m(\tau - kT) - \frac{2\pi n}{T}). \quad (3.5.6)$$

In other words, $A_y(\tau, \nu)$ consists of a rectangular grid of shifted τ slices of $A_{h_0}(\tau, \nu)$ that have been sheared in the ν direction by slope m (see Figure 3.8).

If these terms do not overlap, then we can window $A_y(\tau, \nu)$ about the origin and recover the single term $A_{h_0}(\tau, 0)\delta(\nu - m\tau)$. We can

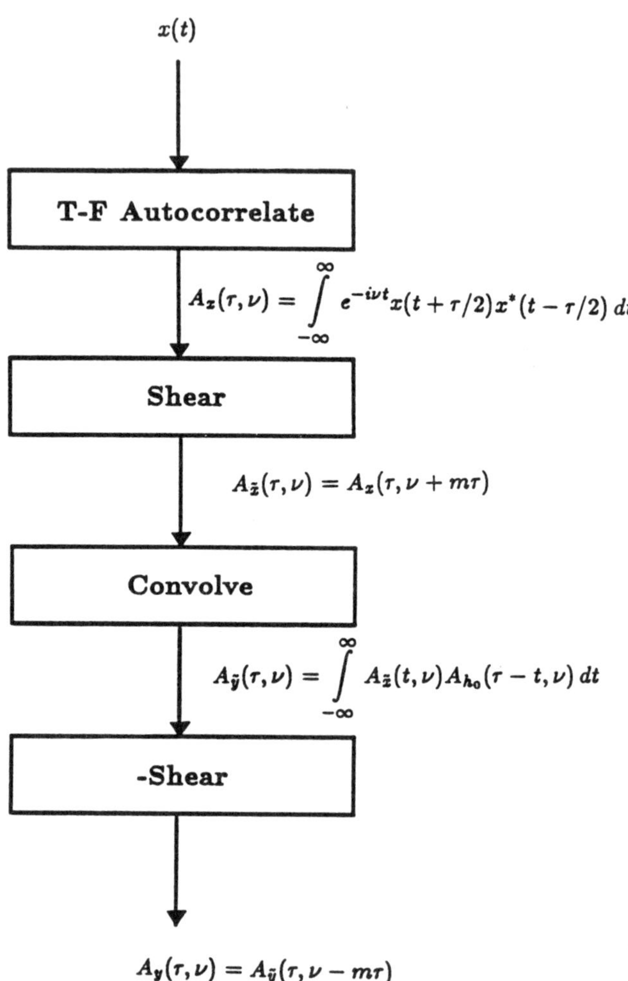

$$A_x(\tau,\nu) = \int_{-\infty}^{\infty} e^{-i\nu t} x(t + \tau/2) x^*(t - \tau/2)\, dt$$

$$A_{\tilde{x}}(\tau,\nu) = A_x(\tau, \nu + m\tau)$$

$$A_{\tilde{y}}(\tau,\nu) = \int_{-\infty}^{\infty} A_{\tilde{x}}(t,\nu) A_{h_0}(\tau - t, \nu)\, dt$$

$$A_y(\tau,\nu) = A_{\tilde{y}}(\tau, \nu - m\tau)$$

Figure 3.7. *Obtaining the time-frequency autocorrelation function, $A_y(t,\omega)$, of a signal $x(t)$ passed through the filter in Eq. 3.5.1.*

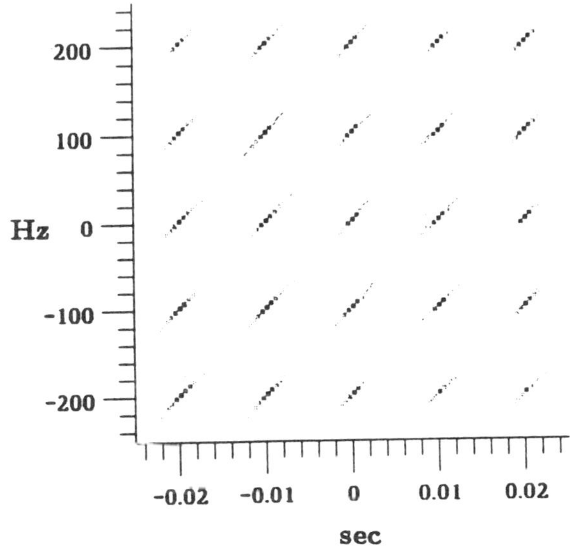

Figure 3.8. *Magnitude of time-frequency autocorrelation function of the output of an FM filter with linearly varying modulation slope (10 Hz/msec) excited by an impulse train (10 msec period). In this example, the corresponding LTI filter consists of a single pole of 300 hz bandwidth.*

then take its inverse 2-D fourier transform to obtain $|H(t,\omega)|^2$:

$$\mathcal{F}^{-1}\left[A_{h_0}(\tau,0)\delta(\nu - m\tau)\right] = \int\limits_{-\infty}^{\infty} \int\limits_{-\infty}^{\infty} A_{h_0}(\tau,0)\delta(\nu - m\tau)e^{i(\nu t - \tau\omega)}\, d\tau\, d\nu$$

$$= |H_0(i(\omega - mt))|^2 \, ,$$

and from Eq. 3.2.3,

$$= |H(t,\omega)|^2 \tag{3.5.7}$$

(see Figure 3.9).

On the other hand, if the terms in Eq. 3.5.7 do overlap somewhat, then a low-pass version of $|H(t,\omega)|^2$ can still be recovered, since

$$\mathcal{F}^{-1}\left[\Phi(\tau,\nu)A_y(\tau,\nu)\right] \approx \mathcal{F}^{-1}\left[\Phi(\tau,\nu)\frac{2\pi}{T}A_{h_0}(\tau,0)\delta(\nu - m\tau)\right]$$

$$= \frac{1}{T}\phi(t,\omega) ** |H_0(i(\omega - mt))|^2$$

$$= \frac{1}{T}\phi(t,\omega) ** |H(t,\omega)|^2 \, , \tag{3.5.8}$$

where $\Phi(\tau,\nu)$ is the time-frequency window, and $\phi(t,\omega)$ is its inverse fourier transform. A 2-D gaussian window is used, and its dimensions are matched to the period T and the fundamental frequency $2\pi/T$, respectively (see Figure 3.10).

So far, we have shown that the time-frequency filtering can be used to estimate the transfer function of two kinds of linear filters — time-invariant and FM filters with linearly varying modulation frequency. We now show that more general cases will follow from the time locality of this operation.

3.6. The quasi-stationary case

We next consider the quasi-stationary case in which the vocal tract changes slowly over time. The traditional way to deal with this situation is to extend the stationary arguments (Section 3.1) by substituting the short-time spectrum for the spectrum of the entire signal. There are thus two windows involved in this analysis; the spec-

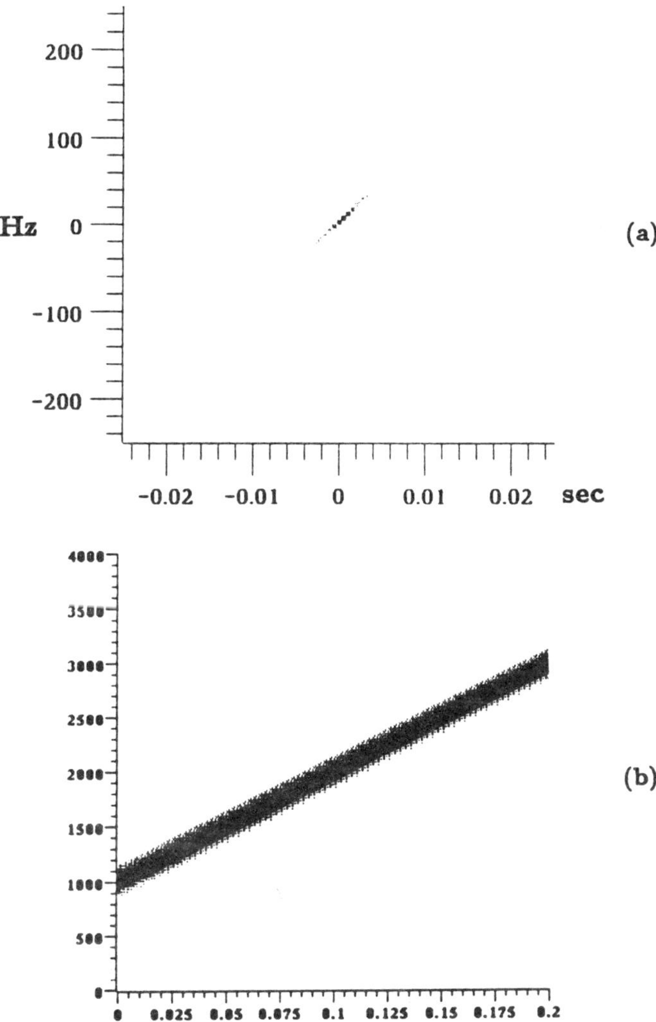

Figure 3.9. *Rectangular window (very nearly) recovers 'unaliased' transfer function.* *(a) Windowed time frequency autocorrelation function in Figure 3.8. (b) Square magnitude of transfer function, the 2-D inverse fourier transform of '(a)'. In the 'aliased' case, i.e., if the terms in Figure 3.8 were to overlap, a gaussian window would be more appropriate.*

trogram window, $w_S(t)$, and the autocorrelation function window, $w_A(\tau)$.

The 'two-dimensional' approach that we have outlined above extends directly without the need of an additional window. In fact, the estimate of $|H(t,\omega)|^2$ is a positive representation of the signal energy

so from Eq. 2.6.5 we know that $|H(t_0,\omega)|^2$ effectively depends only on signal values within a few σ_t of t_0. † Provided the quasi-stationary signal does not change much over this interval, the stationary results of Section 3.4 generalize immediately.

These two approaches for quasi-stationary signals, the former using a 1-D window, $w_S(t)$, on the signal and a 1-D window, $w_A(\tau)$ on the autocorrelation function, and the latter using a single 2-D window, $\Phi(\tau,\nu)$ on the time-frequency autocorrelation function, are related. In fact, $\phi(\tau,\omega) = A_{w_1}(\tau,\nu)w_2(\tau)$. The latter approach specifies the time and frequency scale of interest independently with each of the dimensions of the window $\Phi(\tau,\nu)$, This is somewhat cleaner than the former, which selects the time and frequency scales with its two windows, $w_S(t)$ and $w_A(\tau)$, but not independently.

3.7. Smoothly varying modulation frequency

Suppose the modulation frequency $\gamma(t)$ in Eq. 3.2.2 varies smoothly as a function of time. In other words, it is approximately linear locally, with $\gamma''(t)$ small. For example, a formant with a trajectory that does not have sharp bends in it can be modelled this way. By comparison, quasi-stationarity requires the trajectory have shallow slope, i.e., $\gamma'(t)$ is small.

---

† Provided $\sigma_T \sigma_\Omega \geq \frac{1}{2}$.

The locality argument used in the preceding section to show that the estimate of $|H(t,\omega)|^2$ extends to the quasi-stationary case applies equally to the case here. If the modulation slope, $\gamma'(\tau)$, does not change much over an interval of a few σ_t, then the results of Section 3.5 on filters with a linearly varying modulation frequency generalize immediately to the smoothly varying case. This is because $|H(t,\omega)|^2$ depends only locally on the signal.

3.8. The vocal tract transfer function

Thus far, we have defined the notion of a frequency modulated filter and its time-varying transfer function, and we have shown how to estimate this transfer function from the output signal, provided the modulation slope varies sufficiently slowly. We did this because we modelled each formant pole as an FM filter. The vocal tract is modelled as a weighted sum of formant poles, i.e., its impulse response is

$$h(t, a) = \sum_{n=1}^{N} [z_n(a)h_n(t, a) + z_n^*(a)h_{-n}(t, a)], \qquad (3.8.1)$$

where $h_n(t, a)$ is the impulse response of each pole, Eq. 3.2.1 (cf. Eq. 3.1.4).

How can we define the transfer function of such a filter? Extending the stationary case (Eq. 3.1.1) would suggest

$$H(t, w) = \sum_{n=1}^{N} [z_n(t)H_n(t, \omega) + z_n^*(t)H_{-n}(t, \omega)]. \qquad (3.8.2)$$

There are two advantages of this definition. First, it is a simple generalizaton of the stationary case; it allows us to think of transfer function of the time-varying vocal tract at a given time t as equivalent to the transfer function of a stationary vocal tract for the current articulatory configuration. Second, we shall show that it can be estimated from the speech signal, by the methods we have already presented, in fact. These two conditions, which we can call abstractly *phonetic relevance* and *computability*, are probably the most important for any representation to satisfy in the analysis of speech. Unfortunately, there is no simple relation between the system's eigenvalues or the time-frequency representations of the input and output signals and this definition of time-varying 'transfer function'. These latter notions just do not generalize well to this time-varying case.

Two facts show that the transfer function in Eq. 3.8.2 can be estimated by the time-frequency filtering technique we have described above. The first specifies the effect of variable gain at the filter output on the transfer function estimate, which is given by Eq. 3.9.4 in the next section. The second specifies the effect of adding the output of two filters together on the transfer function estimate. Suppose that $h(t,\tau) = h_1(t,\tau) + h_2(t,\tau)$ and that $|H_1(t,\omega)||H_2(t,\omega)| = 0$. Then $|H(t,\omega)|^2 = |H_1(t,\omega)|^2 + |H_2(t,\omega)|^2$. In other words, superposition holds provided the transfer functions do not overlap. This last condition means that we must consider only regions where the formants are not too close to each other, as we did in the stationary argument in Section 3.1. [cf. Eq. 3.1.11]. † This relation holds not only for the transfer functions involved, but also for the estimates of the transfer functions given by the time-frequency filtering, since they are positive representations of the signal.

Using these two facts, we have

$$\mathcal{F}^{-1}\left[\Phi(\tau,\nu)A_y(\tau,\nu)\right]$$

$$\approx \sum_{n\in\{-N,...,-1,1,...,N\}} \frac{1}{T}\phi(t,\omega) ** |z(t)|^2|H_n(t,\omega)|^2$$

---

† Of course, formants often come close together, but we ignore such time-frequency regions for simplicity in this argument. A more thorough treatment would try to deal with these regions also.

$$= \frac{1}{T}\phi(t,\omega) ** |H(t,\omega)|^2 \qquad (3.8.3)$$

for the filter in Eq. 3.8.1, as desired.

3.9. The transmission channel

It is convenient at this point to consider the effect of the transmission channel characteristics on the estimate of the transfer function $|H(t,\omega)|^2$. The results will prove useful in the next section. We examine two cases — the transmission channel as an LTI system with impulse response $r(t)$, and the transmission channel having variable gain $z(t)$.

There are two facts about the Wigner distribution that we need [Claasen & Mecklenbräuker 1980a]. If $p(t) = r(t) * y(t)$, then

$$W_p(t,\omega) = \int_{-\infty}^{\infty} W_r(\tau,\omega)W_y(t - \tau,\omega)\, d\tau, \qquad (3.9.1)$$

and if $q(t) = z(t)y(t)$, then

$$W_q(t,\omega) = \frac{1}{2\pi} \int_{-\infty}^{\infty} W_z(t,\alpha)W_y(t,\omega - \alpha)\, d\alpha. \qquad (3.9.2)$$

In other words, in the first case the Wigner distributions are convolved in time, and in the second case they are convolved in frequency.

If the spectral shaping of the first transmission channel is gradual, i.e., $r(t)$ is of short duration, then from Eq. 3.9.1, $W_p(t,\omega) \approx |R(i\omega)|^2 W_y(t,\omega)$. If the gain variations of the second transmission channel are slow, then from Eq. 3.9.2, $W_q(t,\omega) \approx |z(t)|^2 W_y(t,\omega)$. It follows from these equations and Eq. 3.5.8 that

$$\mathcal{F}^{-1}\left[\Phi(\tau,\nu)A_p(\tau,\nu)\right] \approx \frac{1}{T}\phi(t,\omega)|R(i\omega)|^2|H(t,\omega)|^2, \quad (3.9.3)$$

and

$$\mathcal{F}^{-1}\left[\Phi(\tau,\nu)A_q(\tau,\nu)\right] \approx \frac{1}{T}\phi(t,\omega)**|z(t)|^2|H(t,\omega)|^2. \quad (3.9.4)$$

Thus, these simple kinds of transmission channels have simple effects of the transfer function estimate. The broadband LTI channel essentially shapes the estimate's frequency slices and the slowly varying gain channel shapes its time slices.

3.10. The excitation

Up to now, we have assumed the filter excitation has been an impulse train. We consider more general (and realistic) forms of excitation in this section.

We can create a general periodic excitation from an impulse train by passing it through a LTI filter whose impulse response $r(t)$ has the excitation's pulse shape. The output can then be passed through the time-varying filter $h(t,a)$. Provided the spectral shaping by $r(t)$ is gradual, i.e., $r(t)$ is of short duration, then these two filtering operations will commute. The assumption is that the time-varying filter can be considered quasi-stationary over the duration of $r(t)$. This is a reasonable assumption for the gradual spectral rolloffs produced in speech excitation. Since these two operations commute under these circumstances, the effect of the filter $r(t)$ on the transfer function estimate is given by Eq. 3.9.3.

Similarly, slowly varying changes in the amplitude $z(t)$ of the excitation will result in corresponding changes in the amplitude of the filter output, with the effect on the transfer function estimate given by Eq. 3.9.4. The pitch period need not be constant, either. Using the locality arguments again, we only require that the pitch period changes slowly.

Finally, consider the case where the filter is noise-excited. Martin & Flandrin [1985] discuss using time-frequency filtering as a general approach for analyzing non-stationary random signals. Our model here involves not only non-stationarity, but also noise that is not additive, and a careful theoretical analysis of this case has not been attempted yet. We must be content, for now, with the following comment. We have seen in the previous chapter that these methods can be used to select time and frequency scales that remove the fine structure introduced by the excitation. This, of course, remains true for this case.

4

The Schematic Spectrogram

4.1. Rationale

In the previous chapters we have seen how to obtain a well-behaved
representation of the the speech energy, with a choice of the time
and frequency scales of interest. For the next step we are faced with
a methodological decision. If we are willing to make strong assumptions about the signal early on, then we can use those constraints
in some detection scheme. For example, one can assume the speech
spectrum is composed of a number of poles, and use analysis-by-
synthesis or linear predictive coding methods to fit these poles to the
spectrum in a formant analysis.

In this approach, a synthetic multiple pole spectrum is fit to each
short-time spectrum. Typically, the pole frequencies can be varied,
but for tractability the number of poles and their bandwidths are
held fixed. Stevens & House [1955] and Olive [1971], for example,
computed mean-square difference between log magnitude short-time

speech spectra and a function of the form:

$$\lg \left| \prod_{n=1}^{N} \frac{1}{(i\omega - s_n)(i\omega - s_n{}^*)} \right| + k, \qquad s_n = \alpha + i\omega_n. \qquad (4.1.1)$$

The poles of the synthetic spectrum that is found to have the least RMS error are taken to be the formants. The permissible range for each of the poles is often restricted to the typical ranges for the corresponding formants in this method. Different versions of this method are identified by the search strategy used to find the best match. Some have used exhaustive search [Stevens & House 1955; Bell, et al 1961; Matthews, et al 1961], so-called analysis-by-synthesis. Olive[1971] used hill-climbing techniques. Linear-predictive coding can be viewed as fitting a fixed number of poles to short-time spectra, using a slightly different spectral distance measure than RMS distance [Atal 1971; Markel & Gray 1976]. The great advantage of LPC is that it provides a simple closed-form solution to the search for an optimum fit.

One problem with this approach, as stated, is that it depends on the quasi-stationary assumption. The short-time spectral contribution of a formant in rapid motion is poorly modelled as a pole with a bandwidth appropriate for a stationary formant. Even when the bandwidths are variable, as in the LPC technique, the diffuse spectral contribution of the moving formant can cause incorrect formant matches. In principle, these methods can be generalized to the time-varying case. Liporace [1975], in fact, has done so for the LPC technique.

This approach, however, suffers from a more general problem. The model used to generate the synthetic spectra has little notion of the source or transmission channel characteristic, or of nasalization.

These effects can contribute significantly to the speech spectrum, "competing" for poles that were meant to be fit to the formants, and thus often resulting in pole distributions that have poor correspondence to the formant distribution. The degree of the fit to a particular point in the spectrum depends on the entire pole distribution; i.e., on the number of poles used and where each pole is positioned in the spectrum. Thus, errors in one part of the spectrum are propagated to other parts in the very first stage in the analysis

For example, Figure 4.1 shows pole locations found by LPC analysis using the autocorrelation method. The order of the analysis was chosen, as is customary, to allow for two complex poles per 1000 Hz plus 4 poles for matching the overall spectral balance (e.g., 12 pole analysis for 4KHz filtered speech). A hamming window was used of 25 msec duration, also a typical choice. In Figure 4.1a, we see that this analysis can perform poorly in regions of rapid formant motion. In Figure 4.1b,c, it appears that the addition of a nasal resonance in the neighborhood of F1 resulted in spurious, unstable behavior in the neighborhood of F3. Decreasing the duration of the window sometimes gives better performance in non-stationary situations, but increases the overall instability of the solution.

The problem, in general, with making such strong assumptions early on in the analysis is that they are seldom universally true. The excitation, the nasal tract, and the transmission channel (e.g. room acoustics and noise) all conspire to make formant analysis more difficult than just fitting poles to a spectrum.

The approach we take here is more conservative, influenced by a similar methodology applied to vision by Marr[1982]. He suggested (1) the *principle of least commitment:* make no decisions that may have to be taken back, later in the analysis, and (2) the *principle of explicit naming:* produce as rich and useful a symbolic description of

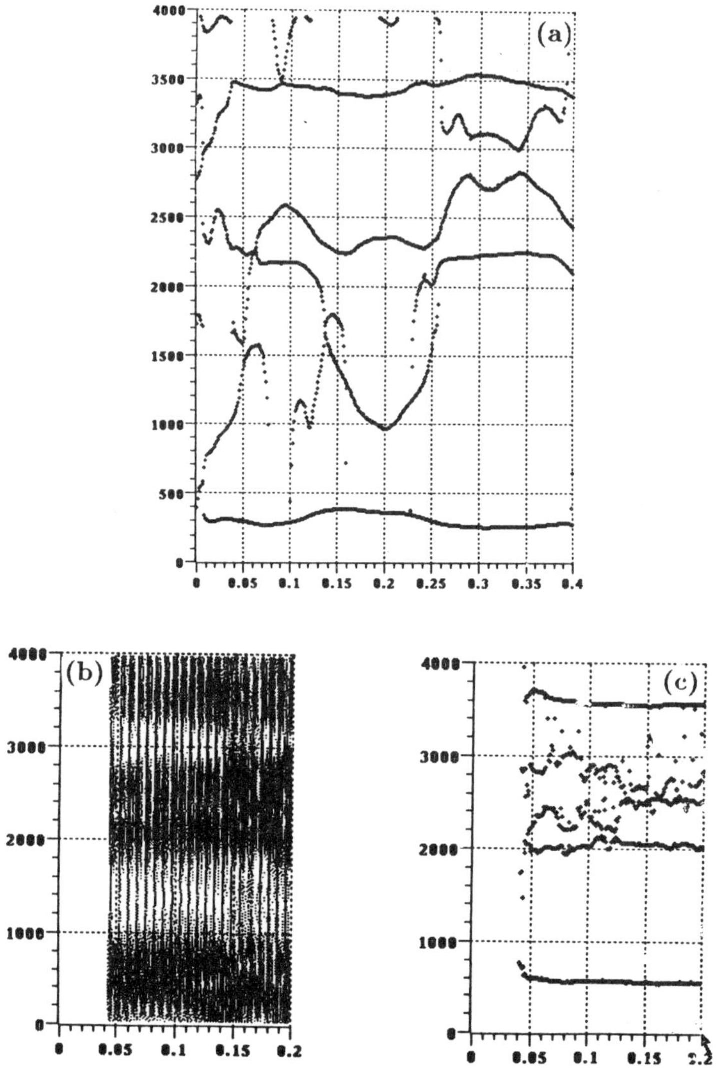

Figure 4.1. *Examples of problems with 'pole-fitting' approach. (a) Poles locations for utterance /wioi/ of Section 2.9. Note the poor performance in the regions of rapid F2 motion. (b) Spectrogram of /ɛ/ in the context /ɛn/. (c) Poles locations for this nasalized vowel. Note the spurious behavior in the neighborhood of F3.*

the input signal as possible, but without any early commitment to its physical origin. This description can be then further organized and analyzed with the goal of finding its physical correlates.

Applying these guidelines to speech suggests taking the energy representations as in Figure 2.15, and producing rich, symbolic descriptions of the significant features there. There are several features (at various scales) that suggest themselves: *time discontinuites* (up and down edges) useful for finding onsets, offsets and bursts; *time-frequency ridges*, easily seen in Figure 2.15, useful for finding the formants and perhaps channel resonances; and some form of *gross spectral balance measure*, also useful for formant and channel analysis. We call this composite symbolic representation the *schematic spectrogram*.

4.2. Spectral Peaks

To create this representation, we must come up with computations that identify these features. This is not as easy as it may seem, since the features clearly visible in Figure 2.15 may nevertheless require some non-trivial computations to detect reliably. We focus on how to find the time-frequency ridges, due primarily to the formants, in the next sections.

An obvious way to try to find these ridges is to identify peaks in vertical slices of the time-frequency energy surfaces. This approach has been tried by several authors, with the main difference between the various instances being how the smoothing was accomplished. Flanagan [1956] used a filter bank whose output was low-pass filtered, Schafer&Rabiner used cepstral smoothing [Oppenheim 1969; Oppenheim & Shafer 1975], while McCandless [1974] used LPC-based smoothing [Atal 1971; Markel & Gray 1976].

To examine this technique, we will use the smoothed time-frequency surfaces of Chapters 2 and 3. Since these surfaces are smooth, the spectral peaks can be found by looking for maxima, i.e., (negative) zero-crossings in $\frac{\partial}{\partial\omega}F(t,\omega)$. Figure 4.2 show these points for the time-frequency energy surface in Figure 2.15. While the horizontal ridge due to F1 is well captured, the steeply rising F2 is very poorly captured. This may seem suprising at first, but the reason is simple.

Eq. 3.5.8 models the situation with F2. The formant pole $H_0[i(w - mt)]$ with time-frequency slope m is smoothed by the 2-D gaussian $\phi(t,\omega)$ to give $F(t,\omega)$. This will produce a time-frequency ridge in $F(t,\omega)$ that has a roughly constant width, independent of slope m, when measured *perpendicular* to the formant trajectory in the time-frequency plane. However, the width of the ridge in a *vertical* slice increases with increasing slope; evidently in Figure 2.15, F2 was sufficiently broadened that its spectral peak is completely lost to other effects in the signal, i.e., other formants, noise, the source and transmission channel characteristic (cf. Figure 2.4).

This effect is not an idiosyncrasy of our particular choice of time-frequency energy representation. It is true, for example, of any representation computed with signal windows (e.g., any positive representation, by Thm. A), since if the formant moves enough in frequency over the duration of the window, its spectral representation will be significantly broadened.

One could rethink the design choices for the time-frequency energy representation, trying for better spectral resolution at the expense of our chosen criteria. However, the problem is not there, as a re-examination of Figure 2.15 will show. The F2 ridge is clearly visible in this representation, it looks no more broadened than the stationary F1. This is because we see both dimensions of time and frequency simultaneously, and as the formant ridge broadens in frequency with

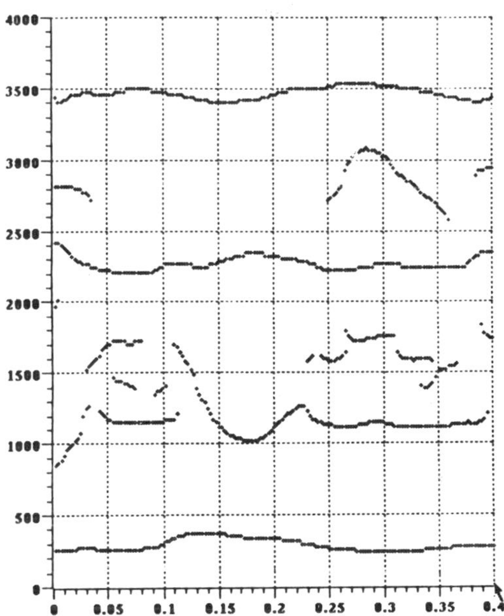

Figure 4.2. *Peaks in spectral cross-sections of the time-frequency energy surface in Figure 2.15. The energy ridge due to F2 is poorly captured by this peak computation.*

Increasing slope it narrows in time. Its prominence depends on its width perpendicular to its trajectory, which does not change much with slope.

Why then did we confine our peak detection methods to vertical slices? It was the usual quasi-stationary prejudice of thinking of speech analysis in terms of a family of one-dimensional spectral analyses parameterized by time. Just like the energy representation problem, this problem is inherently two-dimensional and should be treated as such.

4.3. Time-frequency ridges – non-directional kernel

The approach we will use for detecting time-frequency ridges will depend on whether we use an directional or a non-directional kernel for the underlying energy representation. If we use a non-directional kernel, the problem is simpler, so we shall address this first. In this case, we begin with a single time-frequency representation at a given time and frequency scale, as in Figure 2.15, and the problem reduces to finding the ridges in this smooth, two-dimenional surface.

How can we find ridges in a smooth, two-dimensional surface? This becomes a problem in differential geometry. As such, let us look at the gradient and curvature vectors of the surface in the neighborhood of a ridge. Figure 4.3 shows them for the time-frequency surface in Figure 2.15 in the neighborhood of the initial steep F2. In particular, the solid vectors are used to depict the direction of the gradient, ∇F, i.e., the local direction of steepest ascent. The dotted vectors depict the direction of greatest downward curvature, gdc F, i.e., the local direction in which the surface curves the most downward from the tangent plane.

A precise definition of gdc F is in order. We will use the second derivative as the measure of curvature — this is sometimes called *unnormalized* curvature. This is used instead of normalized curvature (which has the form $\frac{d^2y}{dx^2}/[1+(\frac{dy}{dx})^2]$ in one dimension) for two reasons. First, it is simpler. Second, unnormalized curvature scales linearly with a change in the amplitude scaling, normalized curvature does

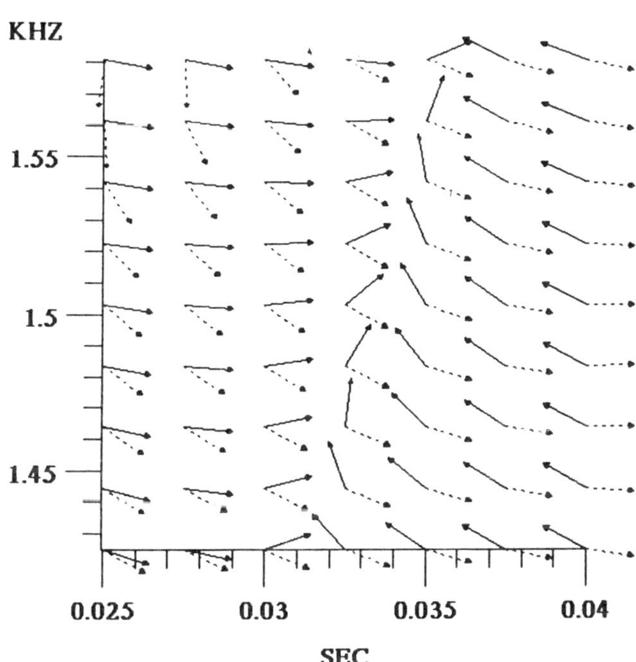

Figure 4.3. *Gradient and curvature vectors in the vicinity of the rising F2 in Figure 2.15. The solid vectors depict the gradient direction, and the dotted vectors depict the direction of greatest downward curvature. (The vector lengths are normalized to unity.)*

not. If wo use the former, our ridge computation proves invariant under changes in the amplitude scaling.

Given this, we define gdc F as the direction vector of the minimum second directional derivative at a given point. More formally, let

$$H(t,f) = \begin{pmatrix} \frac{\partial^2 F}{\partial t^2} & \frac{\partial^2 F}{\partial t \partial f} \\ \frac{\partial^2 F}{\partial t \partial f} & \frac{\partial^2 F}{\partial f^2} \end{pmatrix} \tag{4.3.1}$$

denote the Hessian matrix for $F(t,f)$. Let ξ denote the eigenvector of H corresponding to the lesser eigenvector κ. Then gdc $F = \xi/|\xi|$.

Let us now return to Figure 4.3. As one might expect, the gradient points toward the top the the ridge on each side of it, but must swing through it as one passes over the top. The direction of greatest downward curvature, however, points perpendicular to the ridge in its entire neighborhood, since a surface will curve downward more sharply as one moves toward and away from the top of a ridge then if one moves along it. Note that the two kinds of vectors will become perpendicular precisely on the top of the ridge.

We define the ridge top as the locus of points that satisfy

$$\nabla F \cdot \text{gdc}\, F = 0 \quad \text{and} \quad \kappa < 0, \tag{4.3.2}$$

where κ is the minimum second directional derivative. The inner product of these vectors is zero precisely when they are perpendicular, and $\kappa < 0$ insures that the point is a ridge top and not a trough bottom.

We now show this definition is equivalent to moving along lines of curvature on $F(t,f)$ corresponding to the greatest downward curvature and noting passage through a peak on that surface. This gives an intuitively simple interpretation of a ridge top, and shows that gdc F essentially provides the local ridge direction.

Let $g : \Re \to \Re^2$ be a parameterized, differentiable curve with $g'(s) = \text{gdc}\, F(g(s))$. In other words, g traces out a curve in the time-frequency plane that is always tangent to the direction of maximum downward curvature.   When $F \circ g$ goes through a peak, $\frac{d}{ds}F[g(s)] = 0$.   By the chain rule, this occurs precisely where $\nabla F \cdot g'(s) = \nabla F \cdot \operatorname{gdc} F = 0$.   If $\kappa < 0$, the curve goes through a maximum. † But this is just our ridge top definition, Eq. 4.3.2, as desired.

The inner product in Eq. 4.3.2 is easy to compute for each point on these time-frequency surfaces (one only needs the first and second derivatives of the surface, which are simple to compute for such a smooth surface). Since this quantity may vanish in between sample points in a digital implementation, we detect *zero-crossings* between adjacent sample points.

Figure 4.4 shows the zero crossings in this quantity for the time-frequency energy surface in Figure 2.15. Note that the steep formant peaks are now as well traced as the stationary ones by this ridge top computation. The only thresholding performed here is the removal of points below the signal-to-noise ratio of the analysis. Thus, fairly low amplitude structure can appear in addition to the significant time-frequency ridges. We will examine in Section 4.6 how we to deal with such clutter.

A few pertinent details have not yet been mentioned. First, to perform this computation, an *aspect ratio* has to be chosen between time and frequency, since it is not invariant under different relative scalings of time and frequency. The choice is natural; we use the scaling inherited from the energy representation: let $f = (\sigma_t/\sigma_\omega)\omega$. Thus, we perform our computations in the new co-ordinates, (t, f).

---

† This assumes $|g''(s)|$ is negigible; $(F \circ g)''(s) = g'(s) \cdot Hg'(s) + \nabla F \cdot g''(s)$, where κ equals the first term.

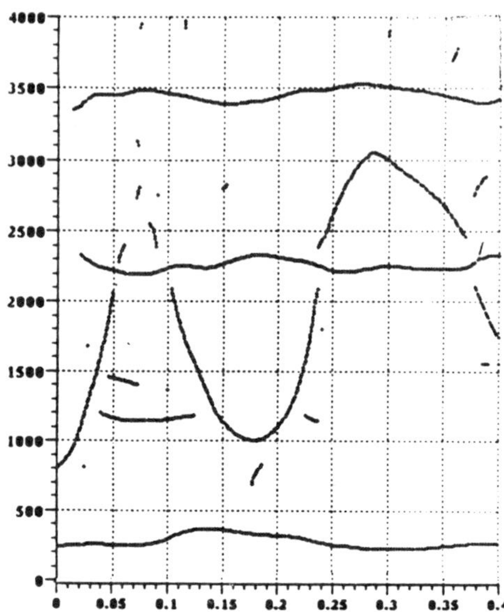

Figure 4.4. *Two-dimensional ridge computation applied to the time-frequency energy surface in Figure 2.15. The contours are those points where the gradient direction and direction of greatest downward curvature are perpendicular. This computation captures the steep time-frequency ridges, due to rapid formant motion, as well as the more horizontal ones.*

Second, very high spatial frequencies have been removed from the energy representation already. Very low spatial frequencies also appear in the vertical direction, due to amplitude variations and formant motion. We find better results when these are also removed by filtering; we thus use a smoothed and *flattened* energy surface for the ridge computation.

4.4. Time-frequency ridges – directional kernel

A second approach to the problem of identifying time-frequency energy ridges uses directional kernels. Let $F(t, f; \theta)$ be a family of time-frequency representations of the class defined by the kernel in Eq. 2.8.8, where θ gives the preferred direction of the transform (i.e., the kernel orientation), and the other free parameters, σ_1 and σ_2, are fixed. We would expect in the vicinity of a time-frequency ridge and for fixed t and f, $F(t, f; \theta)$ would be maximum when θ equalled the local ridge direction θ_0; in other words, when the transform's orientation is tuned to the local direction of the energy ridge. We would also expect that $F[t(s), f(s), \theta_0]$ would be maximum at the ridge top, where $(t(s), f(s))$ is a curve that crosses the ridge perpendicular to its trajectory. The first case corresponds to a maximum under rotation of the kernel; the second case corresponds to a maximum under translation of the kernel along the minor axis of its concentration ellipse (see Figure 4.5).

The locus of points where these two maxima coincide defines a curve in the time-frequency plane, which we can take as our ridge top definition. That is, we seek the points that satisfy both

$$\frac{\partial}{\partial \theta} F(t, f; \theta) = 0 \tag{4.4.1a}$$

and

$$\frac{\partial}{\partial s} F(t, f; \theta) = \nabla F \cdot (sin\,\theta, -cos\,\theta)$$

$$= \frac{\partial F}{\partial t} sin\,\theta - \frac{\partial F}{\partial f} cos\,\theta$$

$$= 0. \tag{4.4.1b}$$

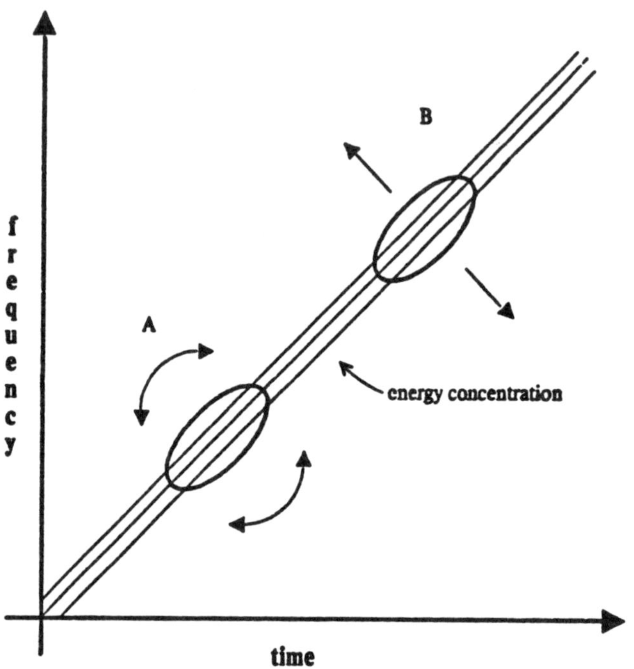

Figure 4.5. *Two conditions for ridge detection: (a) local maximum under kernel rotation, and (b) local maximum under kernel translation along minor axis.*

This computation can be implemented by calculating $\frac{\partial F}{\partial t}$, $\frac{\partial F}{\partial f}$, and $\frac{\partial F}{\partial \theta}$ on a sufficiently fine grid of samples of (t, f, θ), and then finding the simultaneous zero-crossings in the lefthand sides of Eq. 4.4.1a and Eq. 4.4.1b. (The signs of the zero-crossings have to be examined to insure that we have maxima and not minima.)

We yet have to specify the scale parameters σ_1 and σ_2. Alternatively, we can specify σ_2 and $r = \sigma_1/\sigma_2$. We can interpret σ_2 as the size parameter and r as an eccentricity parameter, since the greater the value of r, the greater the eccentricity of the concentration ellipse for the kernel (when holding σ_2 constant).

The choice of r depends on a tradeoff. Clearly, as r increases, time-frequency locality is sacrificed. In particular, bends in the time-frequency trajectory of an energy ridge are poorly resolved with larger values of r.

On the other hand, larger values of r have an advantage in separating intersecting energy ridges, since the larger values of r give better selectivity to a particular orientation. We can quantify this selectivity as follows.

Consider the response of the transform at a frequency f_0 to a complex exponential of frequency f_0. The value is independent of f_0 and equals the value of $F_x(0,0;\theta,r)$ when $x(t) = 1$ (i.e., $f_0 = 0$). We can therefore define a tuning curve $\Gamma(\theta,r) = F_x(0,0;\theta,r)$ that indicates the selectivity of the transform kernel to different values of the orientation and eccentricity parameters.

It is straight-forward to show that

$$\Gamma(\theta,r) \propto \frac{1}{\sqrt{1 + (r^2 - 1)sin^2\theta}}. \tag{4.4.2}$$

In Figure 4.6 this tuning curve is plotted as a function of θ for several values of r.

Even greater orientation selectivity can be obtained if we modify

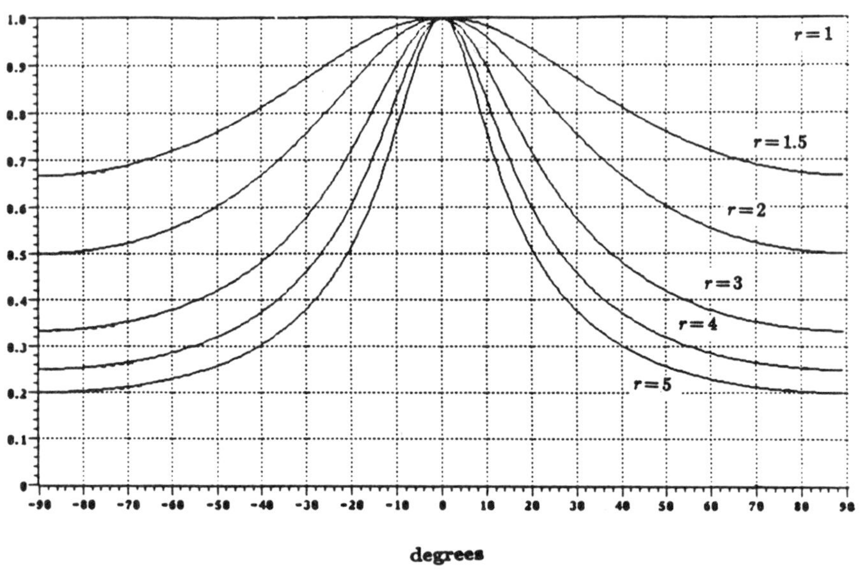

Figure 4.6. *Tuning curves showing directional selectivity of gaussian transform kernels.*

this ridge top computation. The idea is simple; instead of maximizing the energy, $F(t, f; \theta)$, for various θ in Eq. 4.4.1a, we can maximize a more directionally selective measure, such as amount of curvature. In particular, we minimize the second directional derivative perpendicular to the kernel orientation. But this is equivalent to

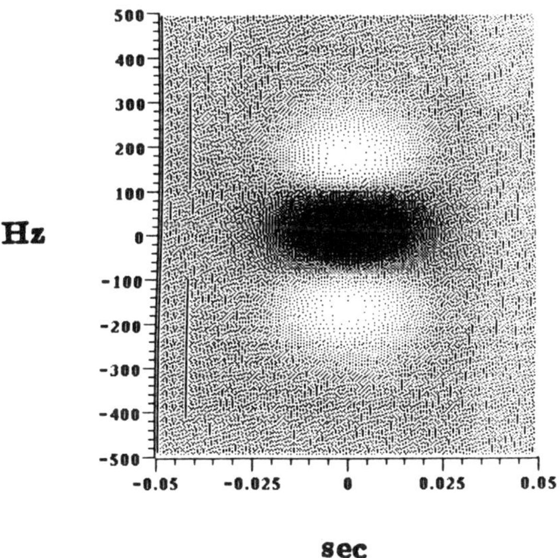

Figure 4.7. Transform kernel $\hat{\phi}(t,f) = -\frac{\partial^2}{\partial f^2}\phi(t,f)$, where $\phi(t,f)$ is a 2-D gaussian. This new kernel has a central 'excitatory' region with 'inhibitory' flanks that give greater orientation selectivity.

maximizing the energy of the transform that uses the modified kernel $\hat{\phi}(t,f) = -\frac{\partial^2}{\partial f^2}\phi(t,f)$; in other words, we use a modified Gaussian kernel in the computation specified by Eqs. 4.4.1a,b. This new kernel has a central 'excitatory' region with 'inhibitory' flanks that give greater orientation selectivity (see Figure 4.7).

The tuning curve for this modified kernel has the form

$$\Gamma(\theta, r) \propto cos^2\theta \, \Gamma^3(\theta, r). \tag{4.4.3}$$

In Figure 4.8 this tuning curve is plotted as a function of θ for several values of r. These indeed show greater selectivity than the corresponding plots in Figure 4.6.

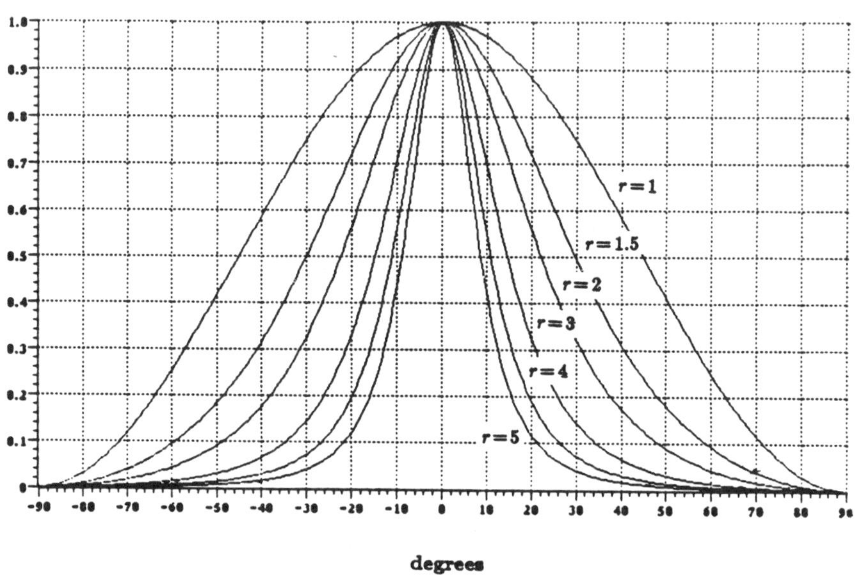

Figure 4.8. *Tuning curves showing directional selectivity of transform kernels of the form in Figure 4.6.*

It turns out that this computation is a generalization of the method in Section 4.3. In particular, if $r = 1$, then the two computations are identical; i.e., those points at which the maximum downward curvature is perpendicular to the gradient direction are identical to those points where the minimum second derivative is parallel to a direction of zero slope.

We therefore see that this section is a generalization of previous section. When $r = 1$, optimal localization in time-frequency results. As r is increased, some of this locality is sacrificed for improved orientation selectivity. Thus, a non-directional kernel will give better results when there is only one ridge in the region, while an directional kernel can give better results when two ridges cross.

Let us examine these results on our example utterance from Section 2.9. For voiced speech, we choose σ_2 to match the pitch period, and we let $r \geq 1$. Then the pitch will be suppressed in each of the $F(t, \omega; \theta)$, using the results of Chapter 3. In Figure 4.9, we show the ridge top analysis on our utterance using the kernel of Figure 4.7 with $r = 2$ and $r = 3$. The case $r = 1$ was shown in Figure 4.4. We see that a less directional kernel (a smaller value of r) gives better performance in the neighborhood of isolated formants, while a more directional kernel (a larger value of r) gives better performance in regions where two formants 'cross' (see Kuhn [1975] for a discussion on the 'crossing' of formants in natural speech.).

4.5. Signal detection and ridge identification

The preceding sections have been based on heuristic arguments. Can ridge identificaton be formulated as a problem in optimal signal detection? We examine this question in this section. Let us begin by making some particularly simple assumptions for ease of argument. We assume that the received 2-D signal representation $F(t, \omega)$ consists of a 2-D deterministic function $S(t, \omega; \gamma(t))$, which depends on the unknown continuous function $\gamma(t)$, plus additive white 2-D Gaussian noise. The problem is to estimate $\gamma(t)$, which models the path of an energy concentration in time-frequency. We further simplify the

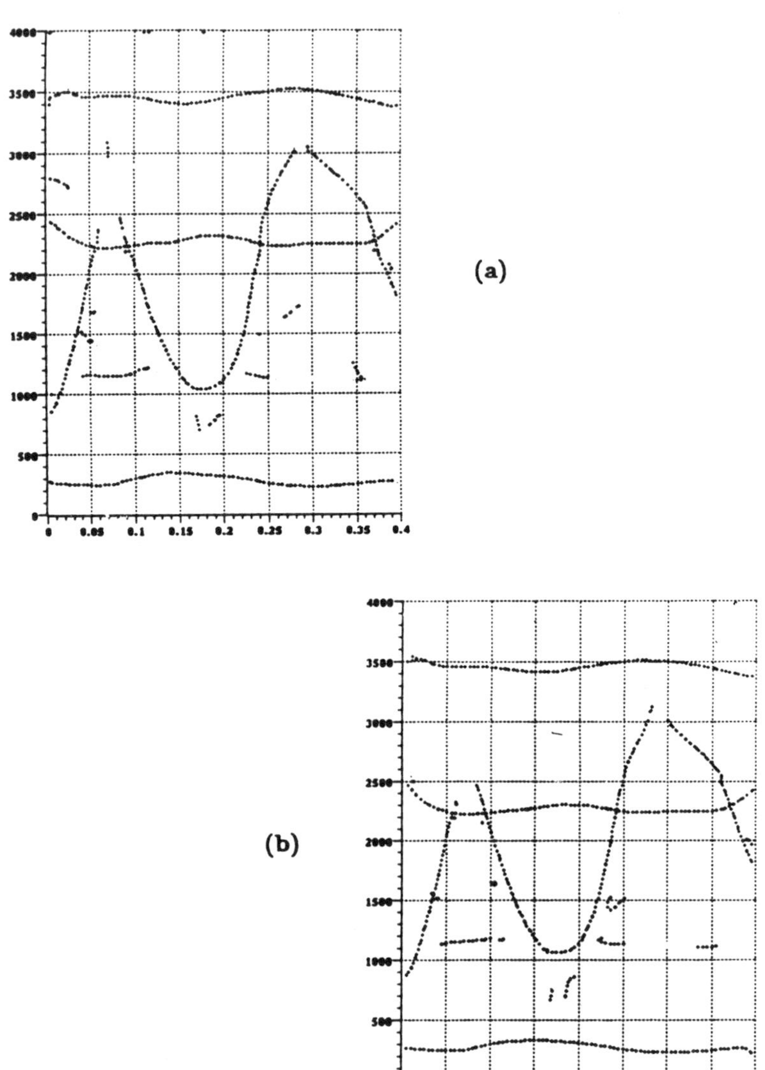

Figure 4.9. *Ridge top analysis of /wioi/ using the directional kernel of Figure 4.7. (a)* $r = 2$. *(b)* $r = 3$. *The more directional kernels give better performance where ridges intersect, but worse peformance at sharp bends.*

problem by assuming that $S(t,\omega)$, which models the energy ridge, has the form

$$S(t,\omega;\gamma(t)) = G(t,\omega) ** \sqrt{1 + [y'(t)]^2}\,\delta(\omega - \gamma(t)). \qquad (4.5.1)$$

In other words, it is a 2-D smoothed (i.e., broadened) curve (the square root factor normalizes the impulse for a unit step in arc length).

In a straight-forward 2-D generalization of the derivation of a matched filter [see Van Trees 1968], the maximum log likelilood estimate of $\gamma(t)$ is proportional to

$$\Lambda[\gamma(t)] = 2 \int\!\!\int F(t,\omega)S(t,\omega;\gamma(t))\,dt\,d\omega - \int\!\!\int [S(t,\omega;\gamma(t))]^2\,dt\,d\omega. \qquad (4.5.2)$$

Substituting Eq. 4.5.1 into Eq. 4.5.2 and changing the order of integration gives

$$\Lambda[\gamma(t)] = 2 \int \sqrt{1 + [y'(t)]^2}\,\hat{F}(t,\gamma(t))\,dt - \int\!\!\int [S(t,\omega;\gamma(t))]^2\,dt\,d\omega, \qquad (4.5.3)$$

where $\hat{F} = F ** G$. The first term is essentially a 2-D matched filter in which the convolution $F ** G$ is matched to the signal shape. The second term takes into account the energy of the deterministic signal. The path $\gamma(t)$ that maximizes Eq. 4.5.3 is the maximum likelihood estimate.

Solving Eq. 4.5.3 for the best path is difficult. In particular, the second term is hard to evaluate (although it is proportional to the arc length of $\gamma(t)$ when it is sufficiently smooth) However, an analysis-by-synthesis procedure could, in principle, be used to compute it numerically. Since we have assumed $\gamma(t)$ is continuous, this becomes a global optimization over t and ω. This is rather like one pole analysis-by-synthesis with a continuity condition imposed on the pole trajectory.

There is a fundamental problem with this approach, similar to the problem with pole-fitting approach discussed in Section 4.1. Because of the non-locality of the optimization, errors at one point can propagate throughout the solution path at this very first stage of the analysis. If the signal were well modelled by Eq. 4.5.3 and the noise well modelled by additive, white Gaussian noise, then this would nevertheless be the best we could do. Realistically, this is not the case. In particular, the "noise" could include a second ridge; one that we shouldn't treat as noise, but as something to detect also. The detection scheme, as formulated, is too global. Instead, we need to make it more local in the time-frequency plane.

Consider a small element Δs of arc length of the curve $\gamma(t)$, which we can rotate and translate in the $t - \omega$ plane. If we hold its position constant, then for sufficiently small Δs, Eq. 4.5.3 will be maximized for that element if it is oriented perpendicular to the direction of greatest downward curvature. If the element's orientation is held constant, Eq. 4.5.3 will be maximized for that element if one translates it in the direction of the gradient. Together these imply that elements aligned on the ridge tops defined by Eq. 4.3.2 will locally maximize Eq. 4.5.3, in the sense that further maximization requires moving along the ridge. These considerations show that the ridge operator of Section 4.3 provides a kind of local solution to the detection problem formulated here.

4.6. Continuity and grouping

We have seen that the ridge detection methods of the previous sections produce piecewise continuous contours. This follows formally from the Implicit Function Theorem; in particular, the zeroes of a continuously differentiable function $f : \mathcal{R}^2 \to \mathcal{R}$ must form continuous contours in $\mathcal{R}^2$. This continuity is a desirable property of the description since it reflects a constraint on the underlying acoustic events that is nearly always valid — loosely, that their spectral content varies (piecewise) continuously as a function of time. For example, formant motion is so constrained. We explore several ramifications of continuity in this section.

First, continuity helps to solve a practical problem in descriptions of this kind. The ridge description, as it stands, can be cluttered with low amplitude peaks unrelated to significant phonetic events. If we try to discard this unwanted structure by setting a threshold, we would have to keep it fairly low, otherwise we could throw out the baby with the bath water, breaking important contours into fragments. Continuity lets us use thresholding with hysteresis, which is often used in such cases [cf. Canny 1983]. The idea is to set *two* thresholds. Points below the lower threshold are first discarded. Points that are above the higher threshold are retained, as are any points between the two thresholds, provided they lie on a contour that crosses the higher threshold. The result is that insignificant points are discarded without fragmenting more important contours. The technique can be quite effective; Figure 4.10 shows an example.

One may argue that any kind of thresholding is a mistake, since unrecoverable errors can be made. Instead, one should simply carry along the relative amplitudes and strengths of the various points in the descriptions, and subsequent processing can take these weights into account. This is, in principle, safer, but pratically it is much harder to think about processing a cluttered, weighted description than one that has been first cleaned up. So that the problem does not become too unwieldy at this stage, it is best for now to proceed with a cleaned up description.

Continuity plays an important role in another problem — labelling. Our goal is to eventually be able to label the points in the descrip-

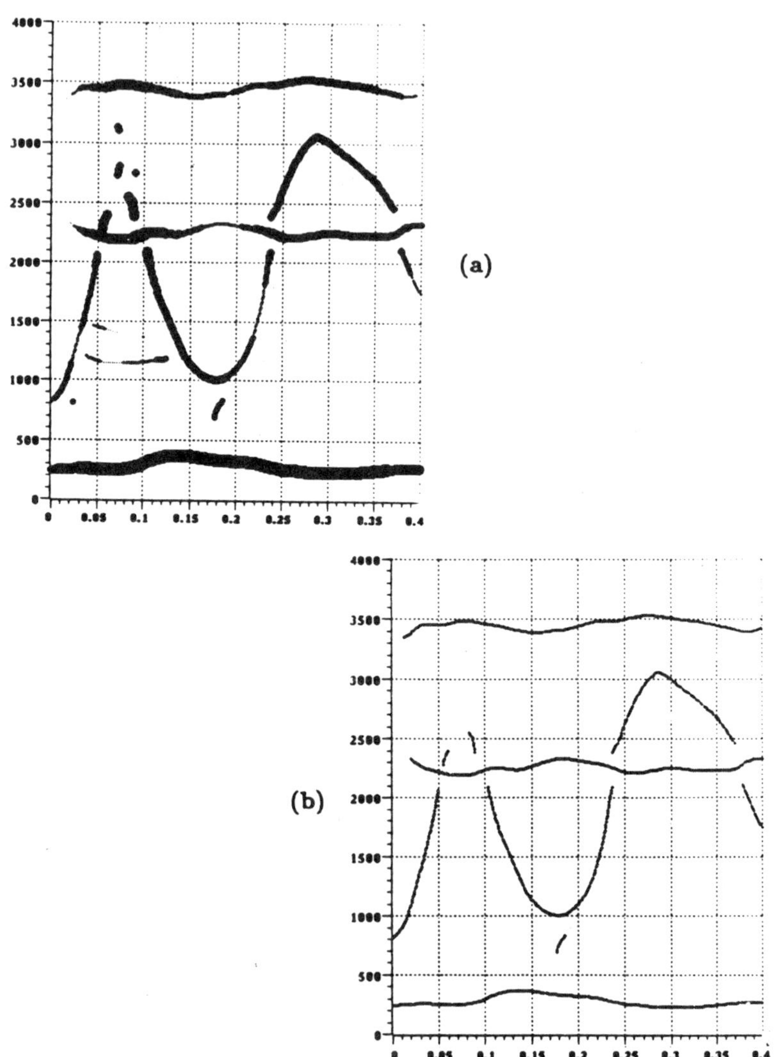

Figure 4.10. *Hysteresis thresholding applied to utterance /wioi/ of Section 2.6. (a) Two-dimensional ridge tops. Amplitude of the ridge top is depicted by the width of the contour. (b) Hysteresis thresholding of '(a)'. This removes isolated, low amplitude points without fragmenting the more significant contours.*

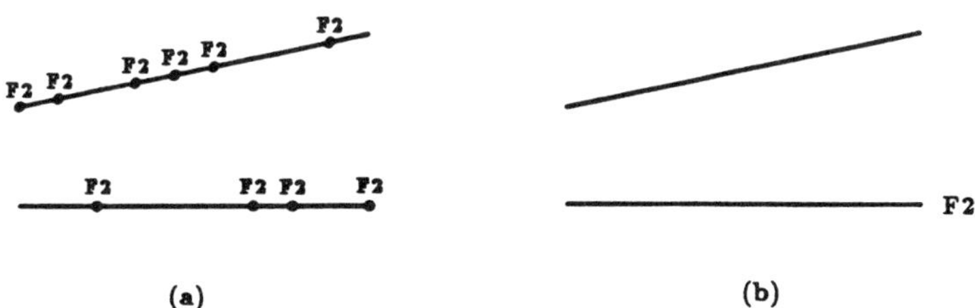

Figure 4.11. *Two contours competing for labelling as F2. (a) One of 2^{10} possible labellings of 50 msec stretch when a new label can be assigned every 5 msec. (b) One of two labellings when whole contours receive a single label.*

tion with their acoustic correlates, e.g., formant identification. This problem would be greatly simplified if a whole contour could receive a single label. For example, suppose points along the two contours in Figure 4.11 are competing for labelling as F2. If the points are sampled every 5 msec, then the points in a 50 msec stretch can be labelled in 2^{10} different ways. If each of the contours, however, is known to have a single acoustic correlate, then there are only two possible labelings.

This is a simple point, but it is almost universally overlooked. The usual approach has been to label individual points in a spectrum, and then either ignore continuity altogether, or use it to narrow the range of candidate labellings after the fact. The latter approach leads to a combinatorial explosion of possible labellings. Algorithms such as dynamic programming can be used to make this approach more manageable, but then the effect of even a single error can be catastrophic. A more direct approach is to first identify stretches of contour that will receive a unique label, with each deemed to have a single acoustic correlate.

How can we identify such "atomic" contours? Ideally, our initial analysis would only return such contours. Acoustic events would never be merged into a single contour, but would always be resolved as separate. I do not believe such a "perfect" analysis is possible. It is evidently possible to fool our auditory system on this account. Consider the spectrum of an /i/ in Figure 4.12a. By low pass filtering, the spectrum can be tilted to appear as in Figure 4.12b. This will be perceived as an /u/; the F1 of the /i/ is taken as both F1 and F2. Conversely, an /u/ can be high-pass filtered to sound like an /i/, with F1+F2 being taken as F1.

Listeners seldom make these kind of mistakes with more natural utterances altered by this kind of filtering. This is because they hear them in context, with continuity being an important contextual cue. For example, consider Figure 4.13, which shows the spectrogram of /wi/. The /i/ in Figure 4.12 was taken from this utterance. If the entire /wi/ is low-pass filtered in the manner of Figure 4.12, it is perceived as /wi/, and not as /wu/. Similarly, a high-pass filtered /yu/ will not sound like it ends in /i/.

There are two points to be learned from these examples. The first is that it is probably not possible to always separate distinct acoustic correlates of nearby energy concentrations *locally*, i.e., they can be

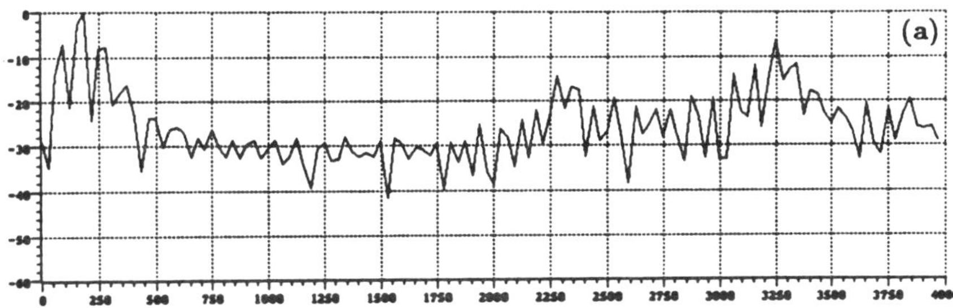

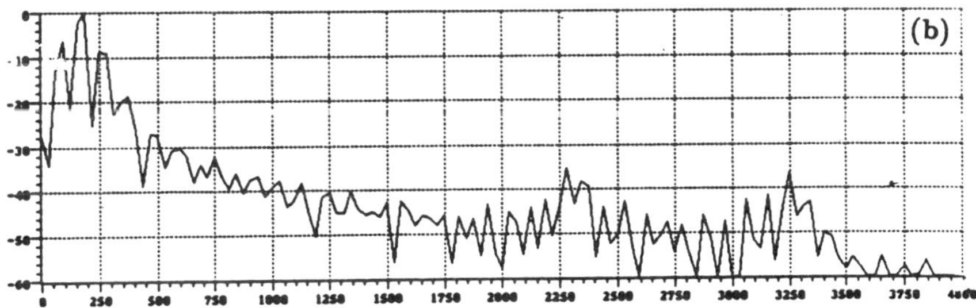

Figure 4.12. *Turning an /i/ into an /u/. (a) Short-time spectrum of an /i/. (b) Low-pass filtered /i/. This will be perceived as an /u/. In other words, F1 is perceived as F1+F2.*

merged if heard in isolation. The second point is that more *global*

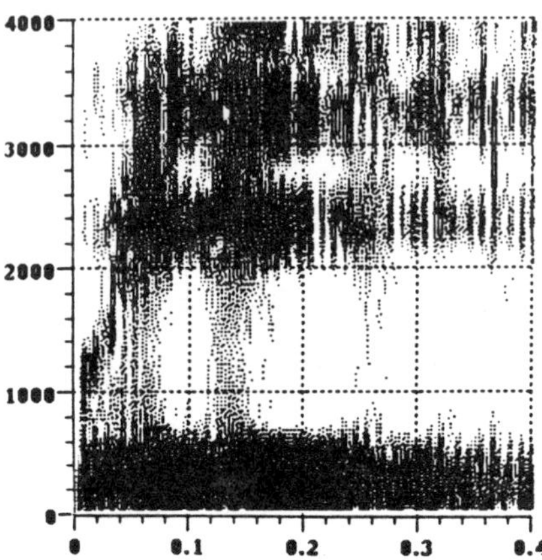

Figure 4.13. *Spectrogram of /wi/. When this utterance is low-pass filtered as in Figure 4.12, it is still perceived as /wi/. Continuity of the formants allows the correct perception.*

constraints, such as continuity, can resolve these mergers.

The ridge description will represent sufficiently close formants with a single ridge, as in Figure 4.14. When the formants merge, one of the contours terminates, and the other continues on. When the formants split, a new contour appears, while the old contour continues on. Evidently, some contours can change their label along their length. For example, the contour in Figure 4.14 that begins as F1+F2 splits into F1 and F2. Obviously, we can not label whole contours with a single label always.

We, can, however, label portions of contours between splits and mergers with a single label. Said differently, if we identify the locations of splits and mergers, we can break the contours into a set of "atomic" contours, in the sense that each contour will receive a single labelling. Since mergers are sparsely distributed in time-frequency, we will still have a small, manageable set of contours.

The idea, then, is to augment our representation to include the locations of splits, mergers, and crossings of contours. Identifying these *junctions* will serve two purposes. First, contour segments away from them can receive single labels along their length. Second, the junction itself can embody continuity constraints, since the junctions must be consistently labelled. For example, if two contours enter a junction and one leaves it , we may label the exiting contour with the union of the labels of the entering contours.

This is somewhat reminiscent of the junction labelling problem in the blocks world. Perhaps an efficient algorithm to propagate these constraints can be found for formant labelling as Waltz [1975] found for the blocks world. The problem here is greatly complicated by the fact that there can be many kinds of errors, e.g., a formant can be "missing". Further, other factors such as spectral balance must be taken into account. We will not attempt any labelling here. Instead, we provide a description of the signal that is a reasonable step toward that goal.

Provided the ridge description is not too cluttered, which is the rule once low amplitude contours have been removed, the identification of contour junctions is relatively easy. In fact, using the proximity of contour endpoints to other contours is a simple method. Two nearby

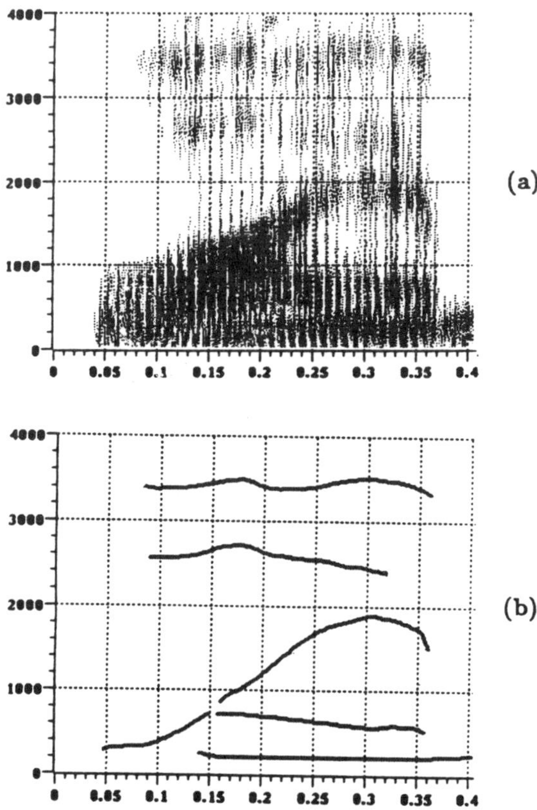

Figure 4.14. *Merged formants. (a) Wideband spectrogram of utterance "why am". (b) Ridge tops. When F1 and F2 approximate, their ridges merge.*

endpoints define a two point junction. Three nearby endpoints or a single endpoint near the body of another contour define a three point junction and so on. Figure 4.15a shows junctions identified by such proximity rules. Contours that both enter and leave a junction are broken there, while two point junctions can be bridged provided that simple "good continuation" rules are satisfied. The result is a set of contours that are likely to have unique labels of their acoustic correlates along their length. Figure 4.15b shows points where contours are broken based on these junctions.

4.7. A perspective

We have shown that the above analysis in some circumstances can produce a more reasonable schematization of the speech signal than, for example, LPC analysis. We will give many more examples of this analysis in the next chapter. Does this mean that the ridge analysis is uniformly better than LPC analysis in speech applications? The answer is no. The simplicity and speed of the LPC algorithms make them attractive for many applications. Further, such pole-fitting models do work well in many cases. Since they embody additional constraints compared to the raw ridge analysis, they will usually not make the 'mistake' of merging nearby formants together. Further, insignificant peaks usually do not affect the pole placements. This means that in clean, unnasalized, quasi-stationary male speech LPC analysis can be quite good. In such cases, the ridge analysis may nevertheless merge nearby formants together and may include additional ridges, making that analysis appear inferior to the LPC analysis.

This probably means that the ridge analysis will offer no improvement in simple speech engineering applications to the widespread LPC methods. Frankly, the power and importance of the ideas presented here comes only when one asks the question: What methods

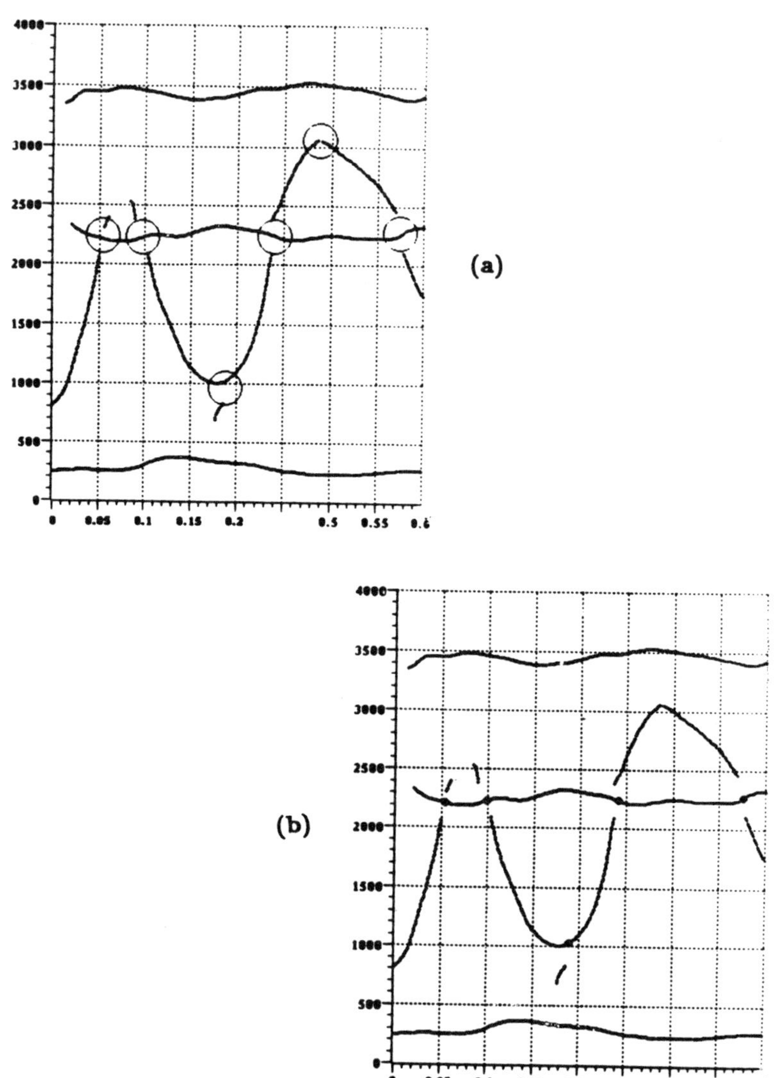

Figure 4.15. *Contour junctions located. (a) Ridge tops of /wioi/ with junctions identified by simple proximity rules. (b) Dots show points where contours are broken based on these junctions.*

will be appropriate for speech analysis in general, natural settings?
Under such circumstances, the transmission channel will often be imperfect and varying (e.g., walking down a hallway with open doors),
there can be environmental sounds and nasalization present, and
there can be significant non-stationarity. In these cases, the very
constraints (i.e., all-pole, quasi-stationary model with a fixed number of poles) that make the LPC technique work so well for 'clean'
speech can cause it to fail in these new circumstances, producing
bizarre pole positionings. On the other hand, the analysis presented
here, a more conservative technique that makes no such assumptions,
will still produce a reasonable schematization of the time-frequency
surface. A simple demonstration of these ideas is given in Section 5.6
below. The key idea is that strong commitments to the origin of the
signal are *not* made at the level of the schematic spectrogram. It is
only after features such as ridge tops, time-frequency edges, temporal discontinuities, and spectral balance information have been made
explicit should articulatory constraints be brought to bear in this
more general, least comittment approach.

5

A Catalog of Examples

In this chapter we will apply the methods of the previous chapters to a variety of examples. This will help us evaluate the strong points as well as the shortcomings of the ideas presented. The ultimate test can come only when these ideas are applied in a recognition scheme. This, however, has not been realized because of the many different components that need to be added, as indicated earlier. At this point, evaluation must be based on any intuitive appeal of the ideas, and on the performance on various examples. Given that the goal is to essentially 'schematize' the information seen in (the sonorant regions of) a spectrogram, an obvious test is to see how reasonable the computed description looks when compared to the spectrogram. Given that previous approaches perform poorly in specific contexts (see Figure 4.1), clear improvements will be apparent.

This situation is similar to edge detection in image analysis. The typical way to evaluate an edge finder is to look at its output compared to the image and ask how good it looks. Perhaps a better test would be to ask how useful an edge finder output is, say, when applied to some scheme for finding surface discontinuities or stereo depth. But such a test requires confidence in the validity of the subsequent processing, since a bad application of a good idea can perform more poorly than a good application of a bad idea.

In Section 5.1, we will look at some general example sentences. In the following sections, we examine several traditional problem categories in speech analysis: in Section 5.2, we look at liquids and glides; Section 5.3 nasalized vowels; in Section 5.4, consonant-vowel transitions; in Section 5.5 female speech. In Section 5.6, we look at some examples of the effects of different transmission channels on the analysis.

5.1 Some general examples

The first four figures of this chapter show the sentences, "May we all learn a yellow lion roar.", "Are we winning yet?", "We were away a year ago.", and "Why am I eager?" spoken by adult males. These sentences were chosen because of their high proportion of sonorant regions and their variety of formant motion. We show wideband spectrograms and the 'ridge' analysis of the previous chapter for each of these utterances. First notice the generally good agreement between the time-frequency ridges seen in the spectrograms and those computed by the ridge analysis; the latter description is a reasonable partial 'sketch' of the former. This is true even in the steeper formant regions, such as the various /w/'s and /j/'s in these examples and at the velar pinch in Figure 5.4 at .75 seconds.

It is important to emphasize that these are not formant tracks, but ridge locations in the time-frequency surface. For example, when two formants come close enough to merge, as in the /wi/ in Figure 5.1 (between .2 and .3 seconds and about 2100 Hz) or a portion of the /r/

in Figure 5.4 (between .85 and .9 seconds and 2000 Hz), only a single ridge is found. (The analysis notes by solid dots the locations that contours should be broken because of possible mergers (cf. Figure 4.15), which can aid in subsequent labelling of the contours.)

There are also ridges present that are not due to the oral formants. For example, the ridge in Figure 5.4 between .15 sec and .55 sec and at about 200 Hz is attributed to nasalization from the /m/. Viewed as a formant tracker this is a failure, but viewed as a ridge detector, this is a success. The nasal resonance is strongly present in the signal in this region and is correctly identified by the analysis. It is properly left to subsequent processing to sort out which ridges are due to formants and which are due to other sources. This is quite different from the LPC analysis, where the presence of nasalization often causes sporadic and bizarre placement of the pole locations (Figure 4.1). In that case, subsequent processing would have difficulty sorting out the situation.

Finally, there are various missing formants. This particularly true for F3 when F2 is quite low as in the /w/ in Figure 5.1. In these circumstances, F3 is driven down by the tail of F2, and is not really visible in the spectrograms either. We know where F3 is by context, but its time-frequency ridge has essentially been driven into the noise.

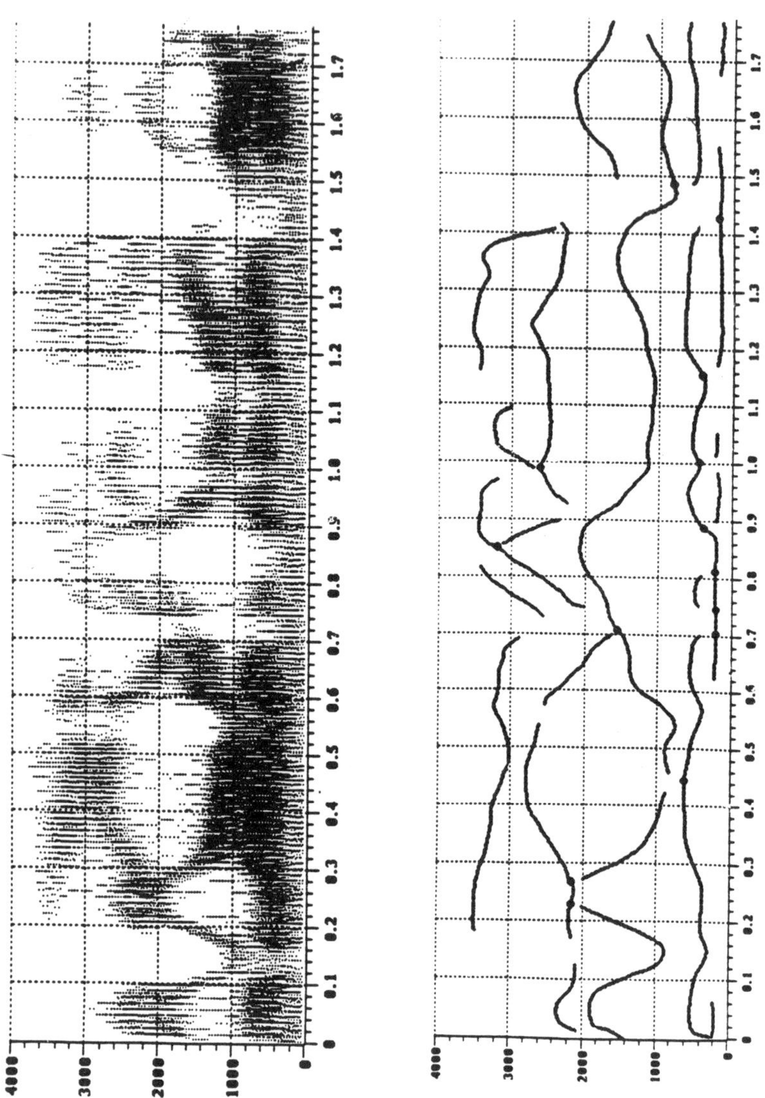

Figure 5.1. *"May we all learn a yellow lion roar."*

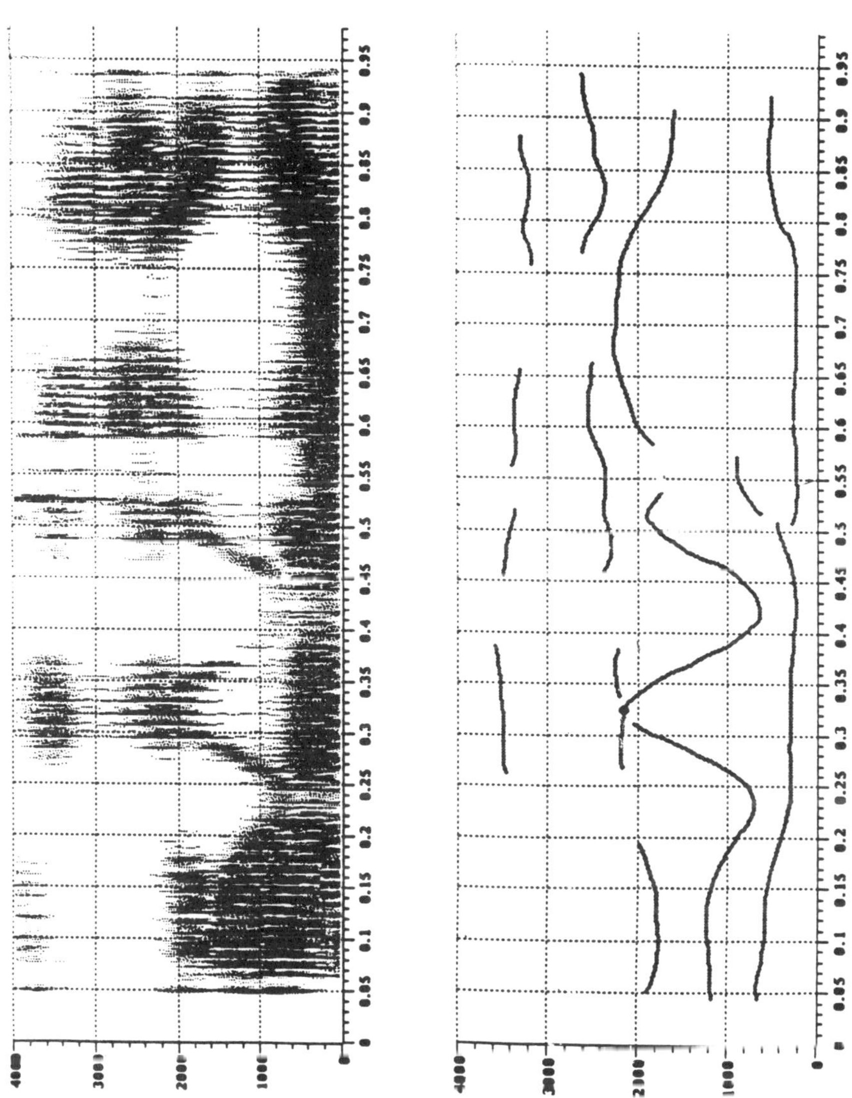

Figure 5.2. *"Are we winning yet?"*

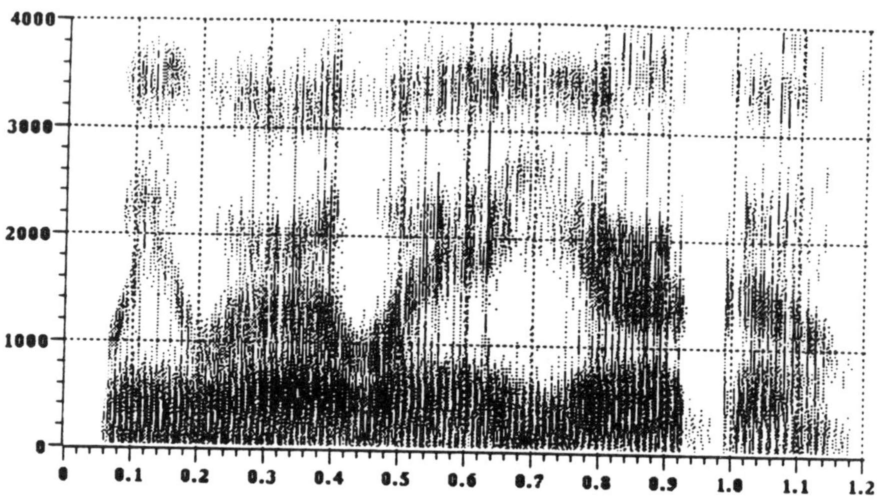

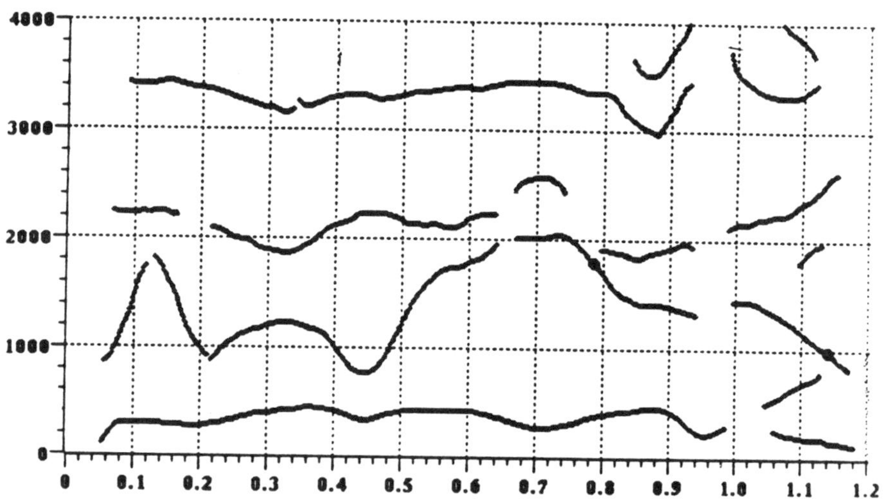

Figure 5.3. *"We were away a year ago."*

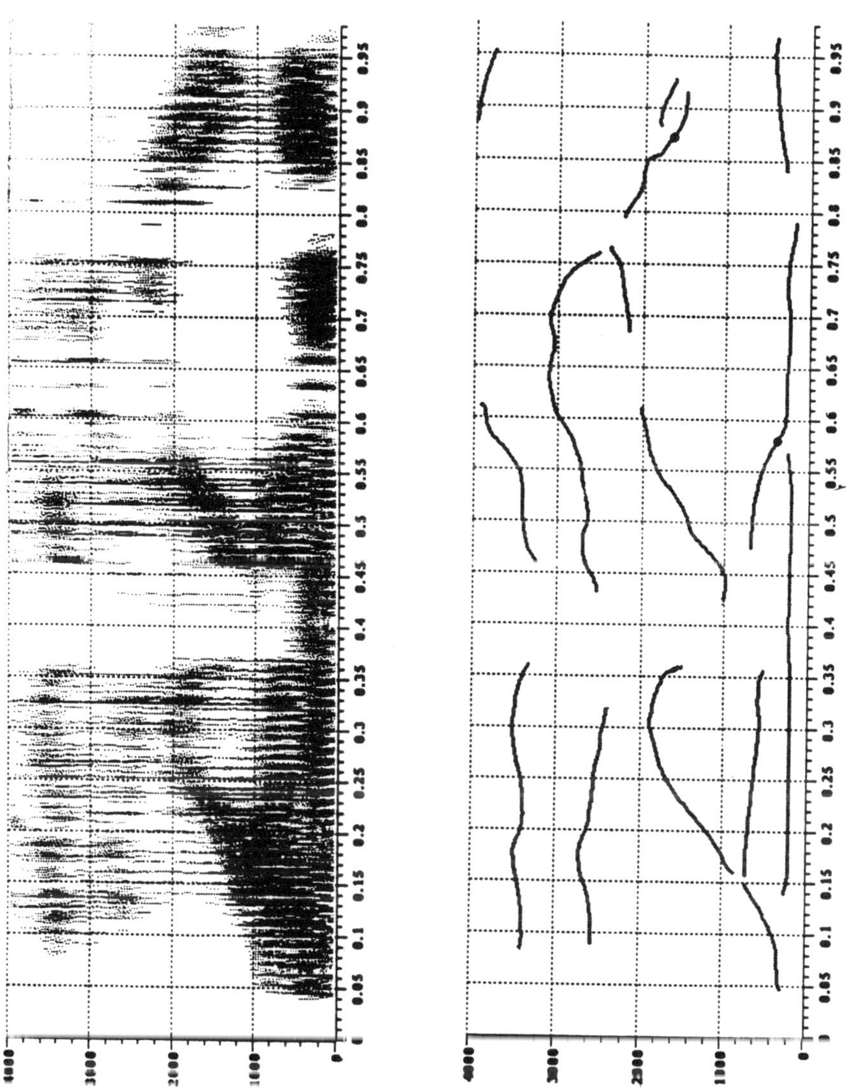

Figure 5.4. *"Why am I eager?"*

5.2. Liquids and glides

In this section we show examples of /w/'s, /j/'s, /r/'s and /l/'s. The /w/'s and /j/'s are syllable initial in the context of /wi/ and /ju/ in Figure 5.5 and Figure 5.6, respectively. A range of speech rates from slow to rapid is shown that gives a range of F2 formant slopes from gradual to steep. Note the ridge analysis is fairly insensitive to this parameter.

The /l/'s in Figure 5.7 are syllable initial, with one example for each of the cardinal vowels, /i/, /ae/, /a/, and /u/. The /r/'s in Figure 5.8 are in the context $V/r/V$, where V ranges over /i/, /ae/, /a/, and /u/. These too show some rapid formant motion that is well captured.

5.3. Nasalized vowels

Figure 5.9 shows syllable initial nasalized vowels in the context $V/n/$. The vowels range over /i/, /ae/, /a/, and /u/. The main feature of this analysis is that additional ridges are introduced due to the nasal 'formants'. As mentioned earlier, this contrasts with the pole-fitting methods, which produce erratic results in nasalized vowels (Figure 4.1).

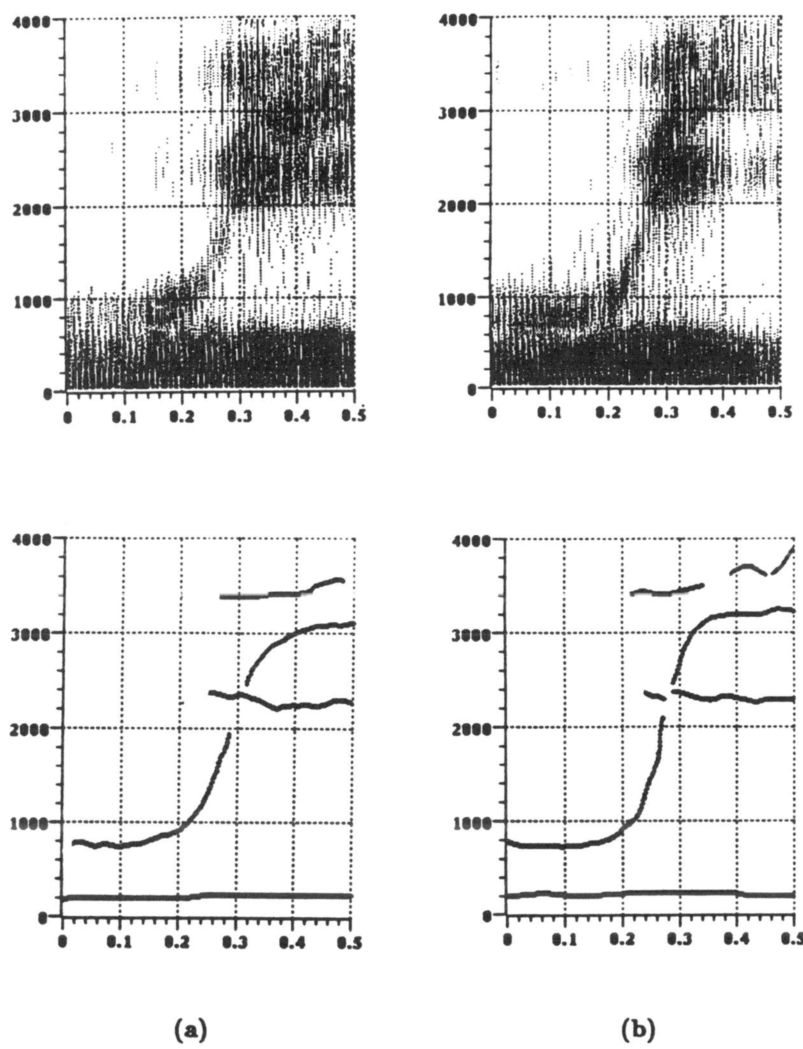

(a)   (b)

Figure 5.5. /w/'s at various speech rates. (a-b) /uwi/. (c-e) Syllable initial /wi/. (continued...)

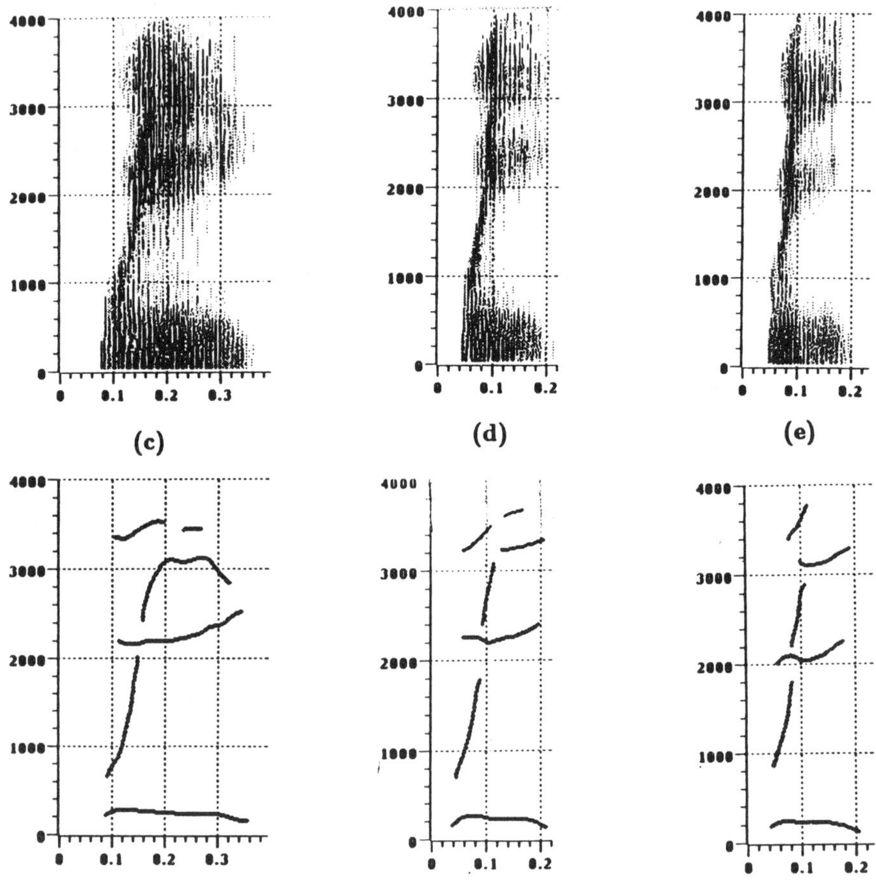

Figure 5.5 (continued). /w/'s at various speech rates. (a-b) /uwi/. (c-e) Syllable initial /wi/.

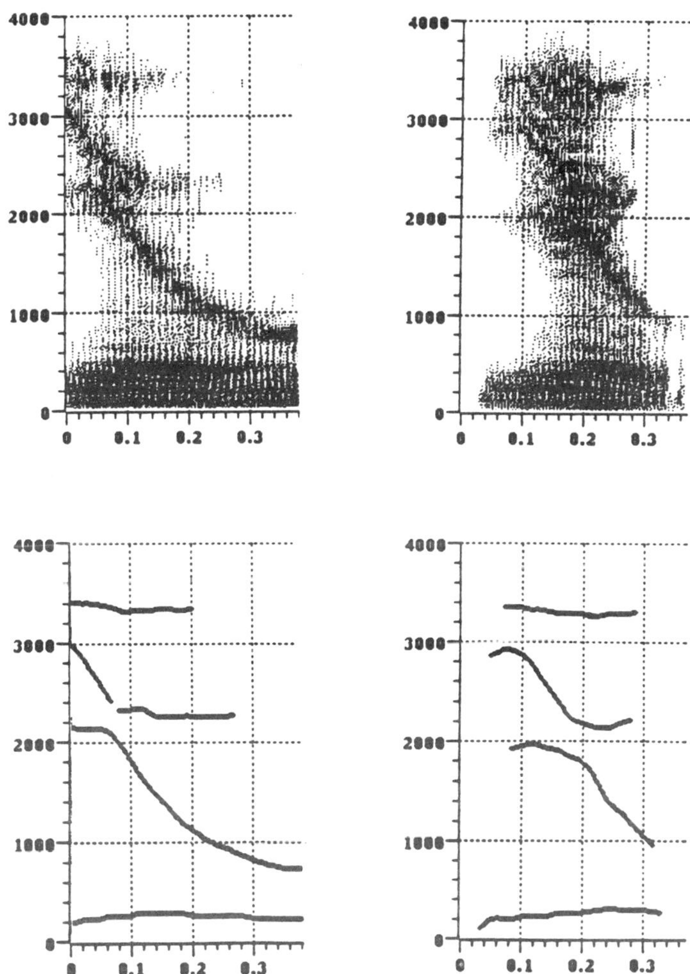

Figure 5.6. *Syllable initial /ju/'s at various speech rates.* (cont'd...)

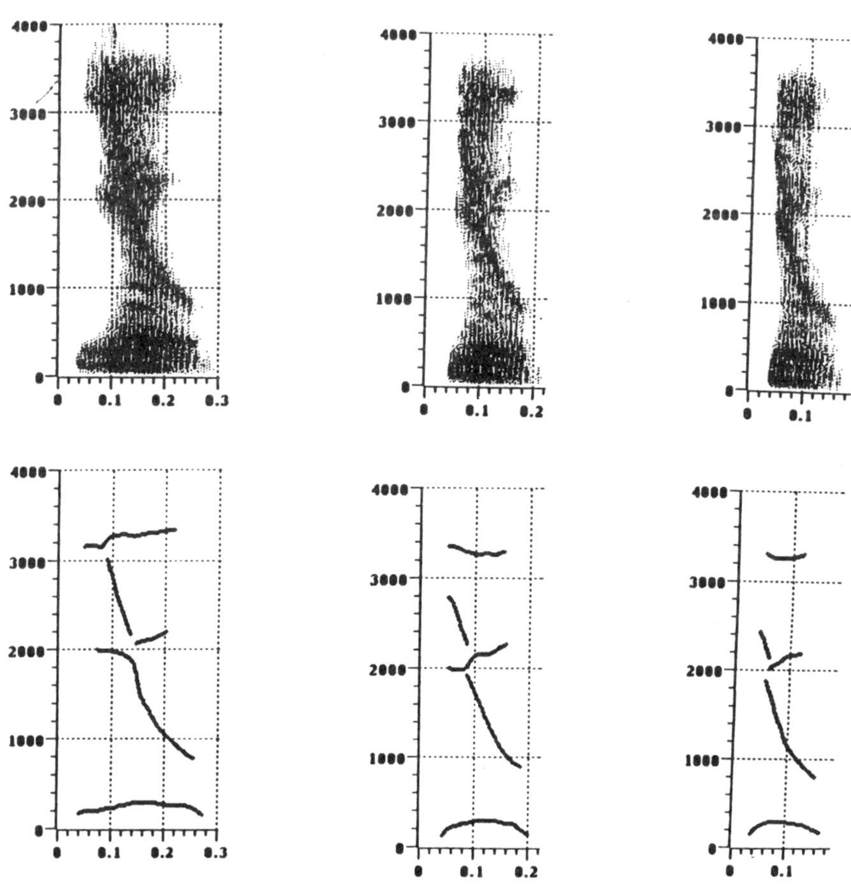

Figure 5.6 (continued). *Syllable initial /ju/'s at various speech rates.*

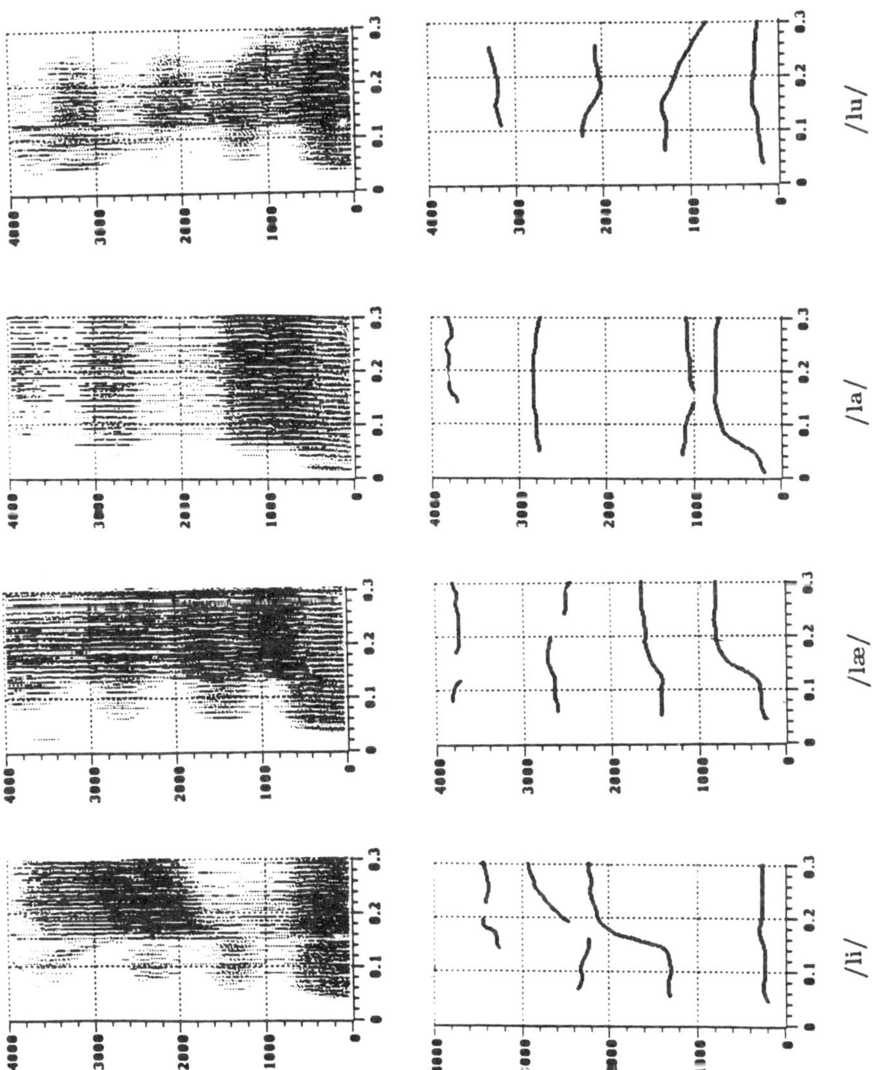

Figure 5.7. *Syllable initial /l/'s.*

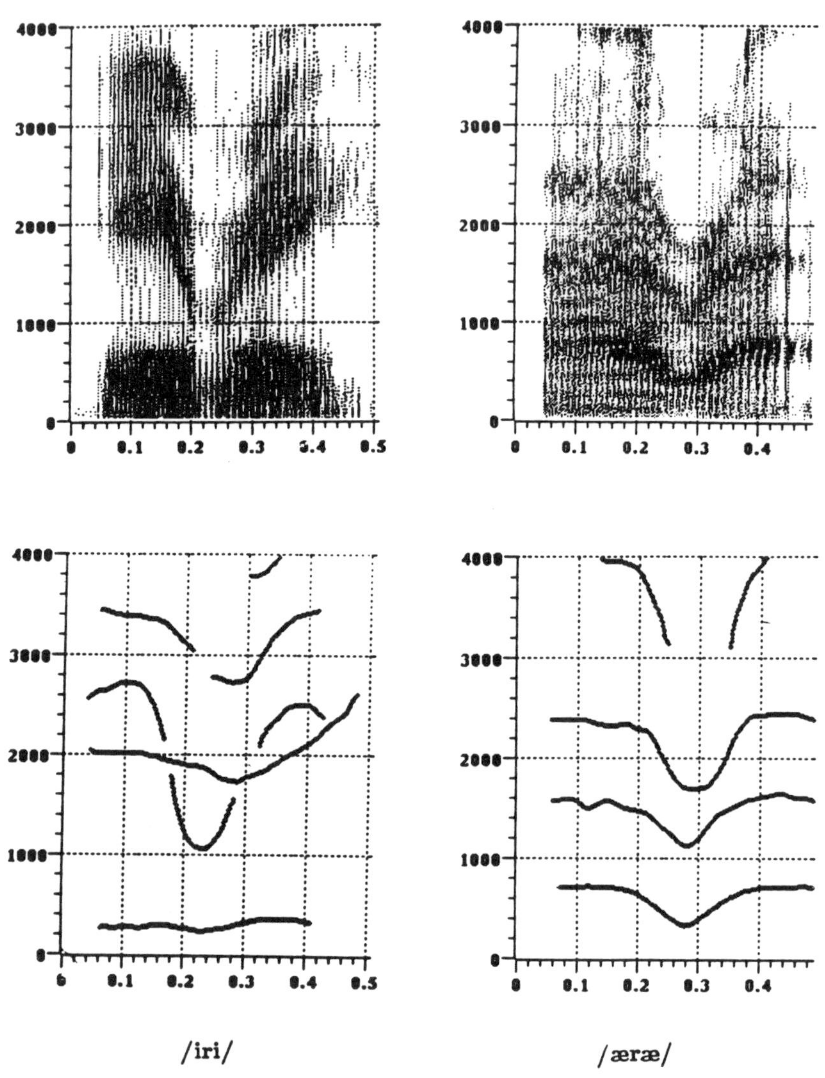

Figure 5.8. /r/'s in various vowel contexts. (cont'd...)

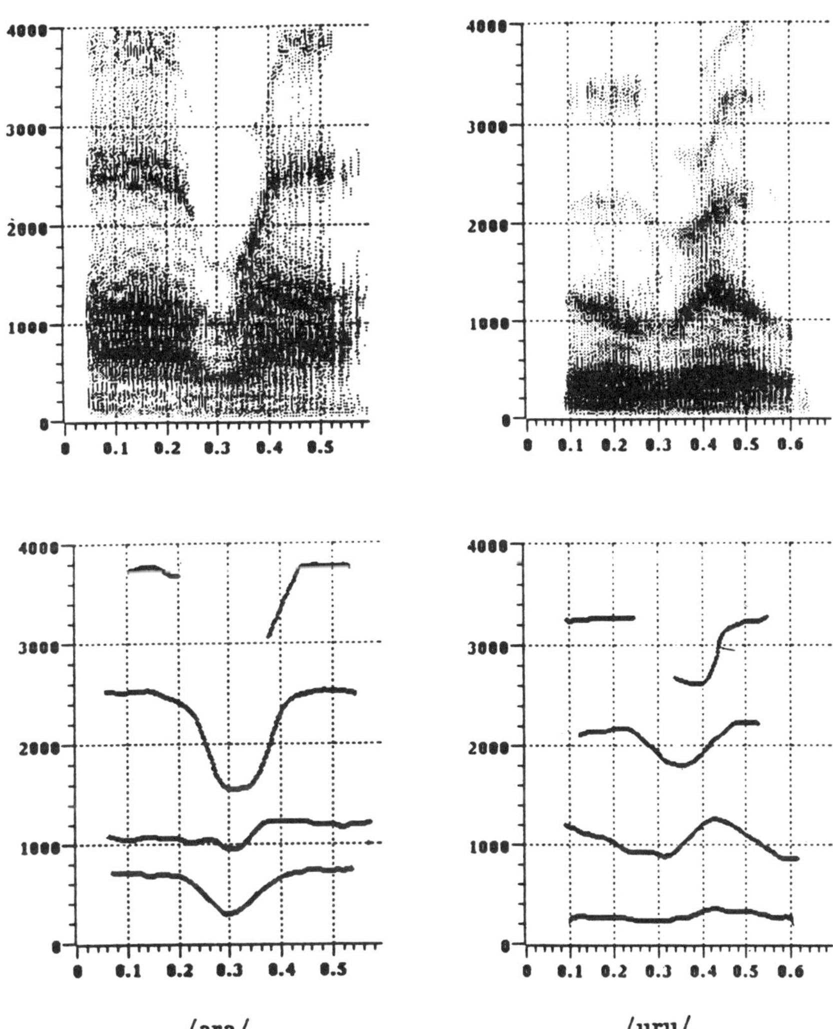

/ara/   /uru/

Figure 5.8 (cont'd). */r/'s in various vowel contexts.*

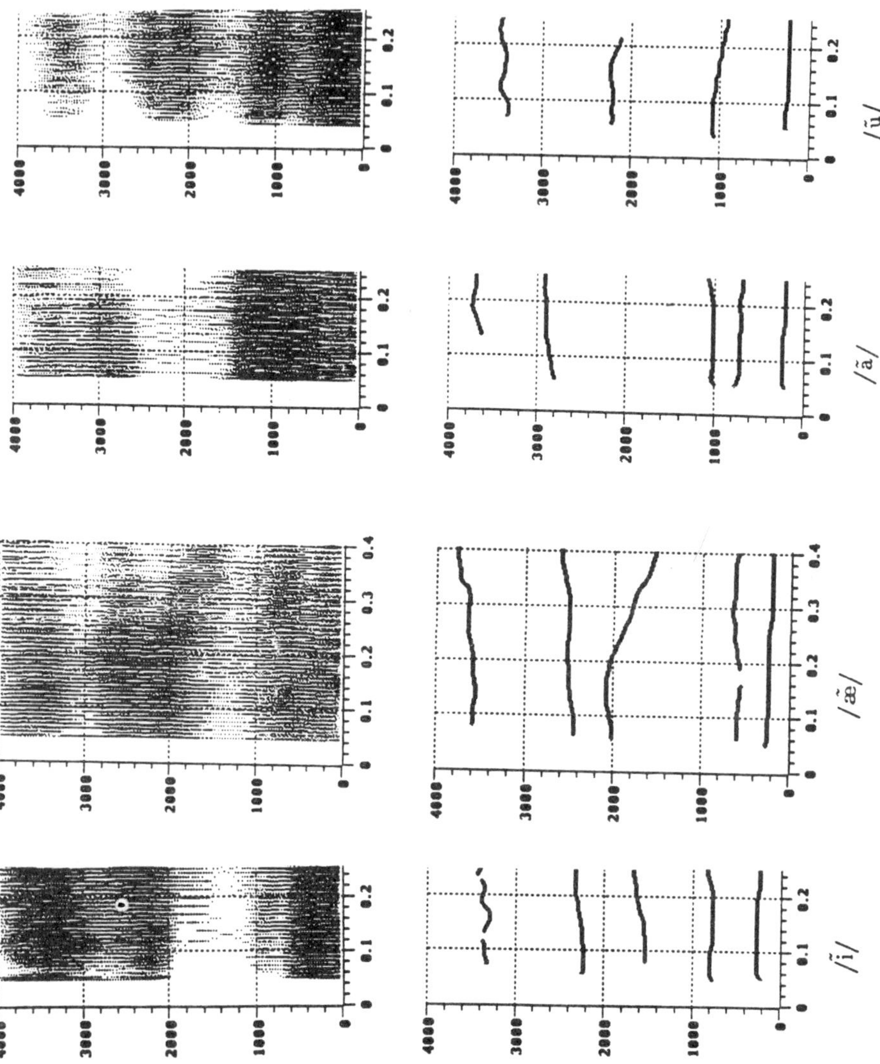

Figure 5.9. *Nasalized vowels.*

5.4. Consonant-vowel transitions

In this section we show examples of consonant-vowel transitions. Figure 5.10 through Figure 5.12 show syllable initial consontant-vowel transitions. The consonants range over the voiced stops /b/, /d/, and /g/ and the vowels range over /i/, /ae/, /a/, and /u/. The analysis is shown only after the consonantal burst since the ridge analysis is inappropriate and peculiar in the burst region. The bursts were located by hand in these examples. Figure 5.13 shows more rapid formant motion with the examples /bi/ in the context /tubi/ and /dw/ in the context /tidw/.

The ridge analysis brings out formant motions consistent with the locus theory of consonant perception. This theory states that one of the cues to the perception of consonants is the trajectories of the formants at the transitions [Liberman, et al 1954]. For example, in many vowel contexts for adult males, F2 will have a trajectory out the consonant that has a locus near about 1200 Hz for labials (e.g., /b/), about 1800 Hz for alveolars (e.g., /d/), and above 2000 Hz for velars (e.g., /g/). This cue is used in spectrogram reading, but has been hard to exploit in automatic speech analysis, because of unreliable formant detection at the often highly non-stationary consonant-vowel transitions.

The analysis here is better behaved, capturing rapid formant ridges as well as shallow ones at the transitions. As noted earlier, however, when the formants approximate a single ridge is produced. The F3 ridge is also sometimes lost near the transition for this speaker; in these cases, F3 appears somewhat diffuse and hard to locate in the spectrograms also. These issues, as well as how to locate the burst, will present difficulties for automatic consonant detection.

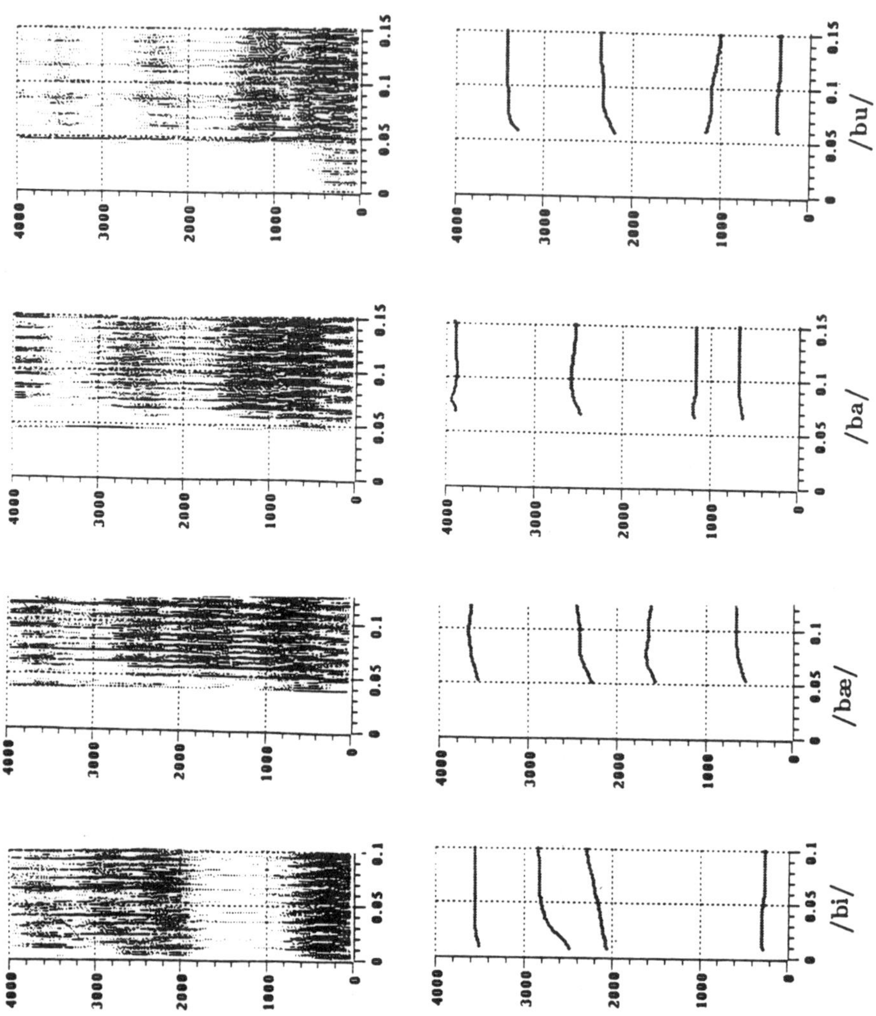

Figure 5.10. *Syllable initial /b/'s.*

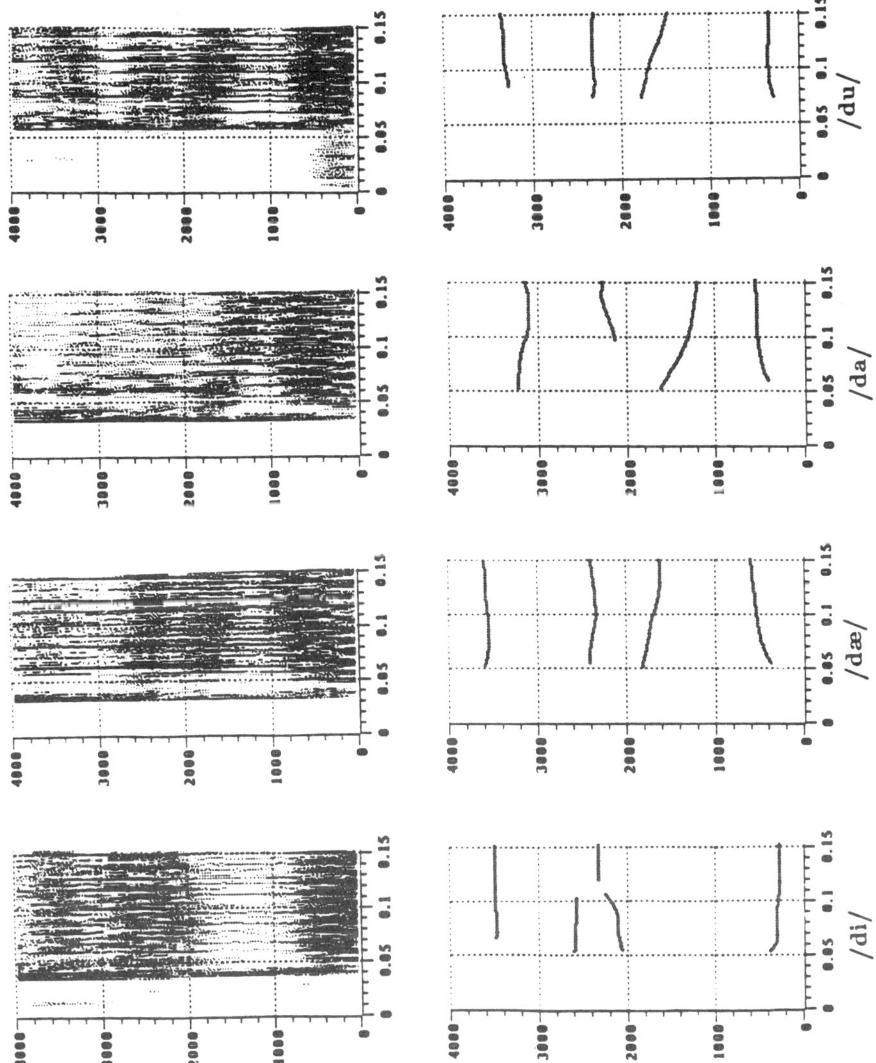

Figure 5.11. *Syllable initial /d/'s.*

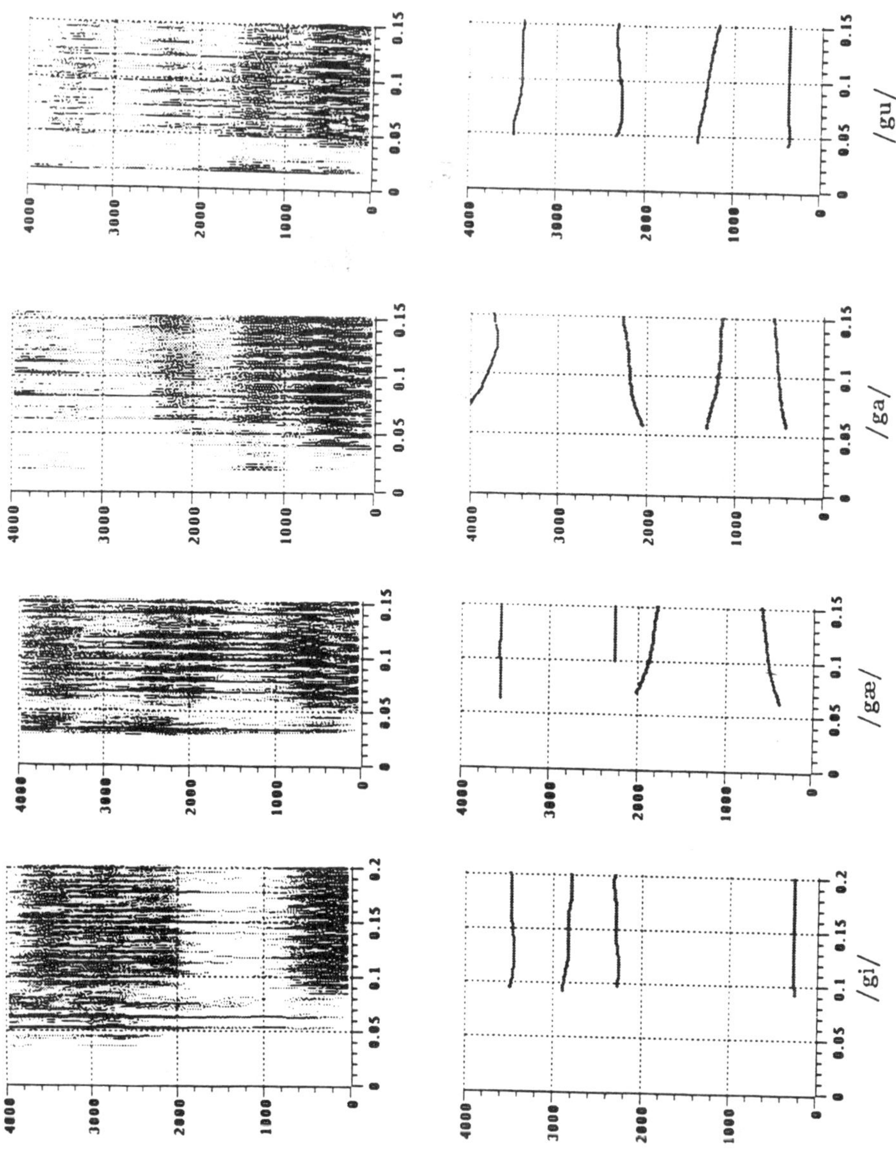

Figure 5.12. *Syllable initial /g/'s.*

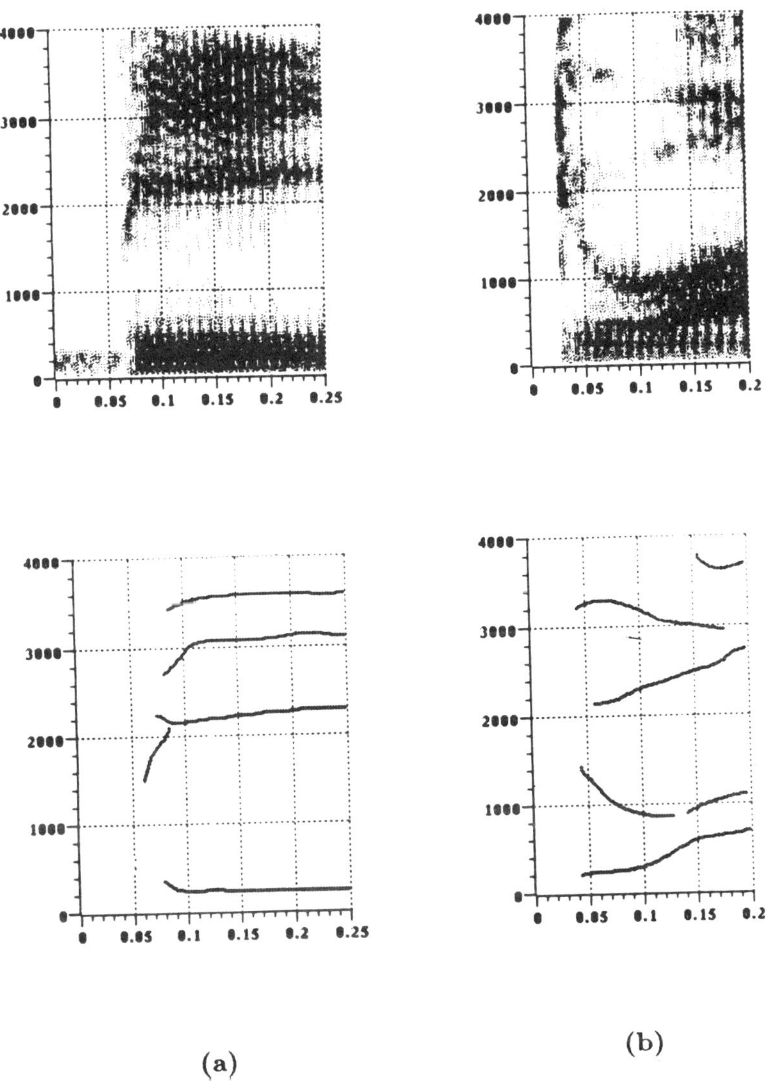

(a)

(b)

Figure 5.13. *Rapid formant transistions. (a) /bi/ in the context /tubi/. (b) /du/ in the context /tidu/.*

5.5. Female speech

Higher pitched speech, such as female and children's speech, present the problem that the harmonics of the (voiced) excitation are fairly widely spaced, viz. a few hundred Hertz or more. This means that in a quasi-stationary analysis, the spectrum is less frequently sampled than for lower pitched speech, resulting in poorer estimates of the vocal tract transfer function (cf. Figure 3.2). Viewed two-dimensionally, the situation is more symmetric. For example, as the frequency of an impulse train is increased, the frequency spacing of the impulses in its time-frequency autocorrelation function (Figure 3.3) will increase, but their time spacing will decrease. Thus one will have poorer frequency 'sampling' of a time-varying transfer function excited by this impulse train, but better time 'sampling'.

The analysis presented in Chapter 3 exploits this fact by matching the time-frequency window to the pitch. Higher pitched speech requires a window at a larger frequency scale but at a lower time scale than lower pitched speech. The remaining analysis proceeds as before. Figure 5.14 gives an example with rapid F2 motion. Figure 5.14a shows a wideband spectrogram of the nonsense utterance /uiuiui/ from an adult female, Figure 5.14b shows the ridge analysis using a time-frequency window matched to a 200 Hz pitch.

Note that the F1 ridge and the steep F2 ridge are well resolved. Where F2 and F3 approximate, however, only a single ridge is found. Such mergers in the analysis are more common in higher pitched speech due to the greater frequency smoothing required. However, since less time smoothing is required than for lower pitched speech, transient effects should, in principle, be better resolved.

5.6 Transmission channel effects

Finally, we consider the effects of imperfect transmission channels on the analysis. In particular, we will consider the effects of passing the speech signal through some simple LTI filters. While the examples we give are idealized, natural environments can give rise to many kinds of transmission channel characteristics. In general, human listeners can tolerate a wide variety of alterations to a speech signal and have it remain intelligible [see Licklider & Miller 1951 for a good review]. That is not to say one is unaware of the modification; e.g., a pronounced room resonance adds a 'hollow' quality to the speech, but it does not destroy its intelligibility.

Figure 5.15 shows the frequency response of the transmission channels we consider. Figure 5.15a consists of a single pole at 1500 Hz of 750 Hz bandwidth, Figure 5.15b consists of a single pole at 1500 Hz of 150 Hz bandwidth, and Figure 5.15c consists of a pole-zero pair – both are at 1500 Hz, the pole has 1000 Hz bandwidth while the zero has 150 Hz bandwidth. Thus, the first channel consists of a fairly broadband, but non-uniform channel; the second channel emphasizes the signal energy in the neighborhood of 1500 Hz; and the third channel removes signal energy in the neighborhood of 1500.

We show the effects of these transmission channels on the analysis of the utterance /wioi/ from Section 2.9. Figure 5.16a shows the wideband spectrogram of this utterance passed through the first channel, and Figure 5.16b shows the corresponding ridge analysis. The effect of this broadband channel is minor when compared to the original analysis in Figure 4.10. Figure 5.17a shows the wideband spectrogram of the utterance passed through the second channel, and Figure 5.17b shows the corresponding ridge analysis. The effect of this narrowband channel is to add an additional ridge at 1500 Hz. Finally, Figure 5.18a shows the wideband spectrogram of the utterance passed through the third channel, and Figure 5.18b shows the corresponding ridge analysis. The effect of this narrowband 'notch' is to put an energy trough in the time-frequency surface, with the F2 ridge being partially cancelled in the vicinity of this notch. Compare this analysis with the LPC analysis of this filtered utterance shown in Figure 5.18c (using the same analysis parameters as in Figure 4.1). We see there that the notch filter plays havoc with the LPC analysis, since the zero lies outside the scope of its all-pole model. This is analogous to the effects of nasalization on LPC analysis.

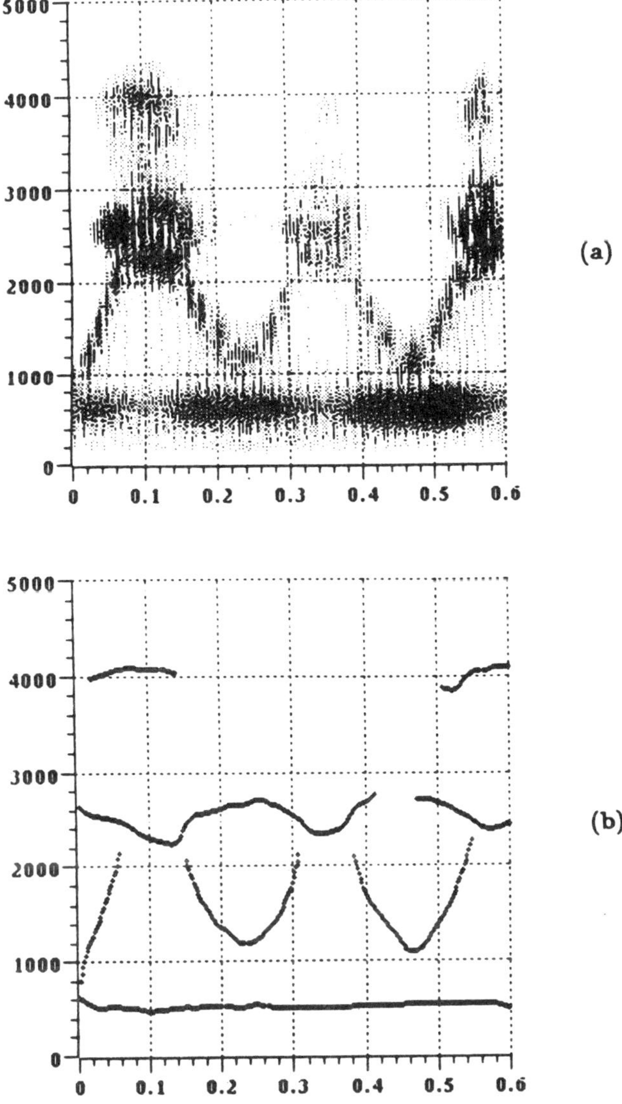

Figure 5.14. */uiuiui/ uttered by an adult female. (a) Wideband spectrogram. (b) Ridge analysis.*

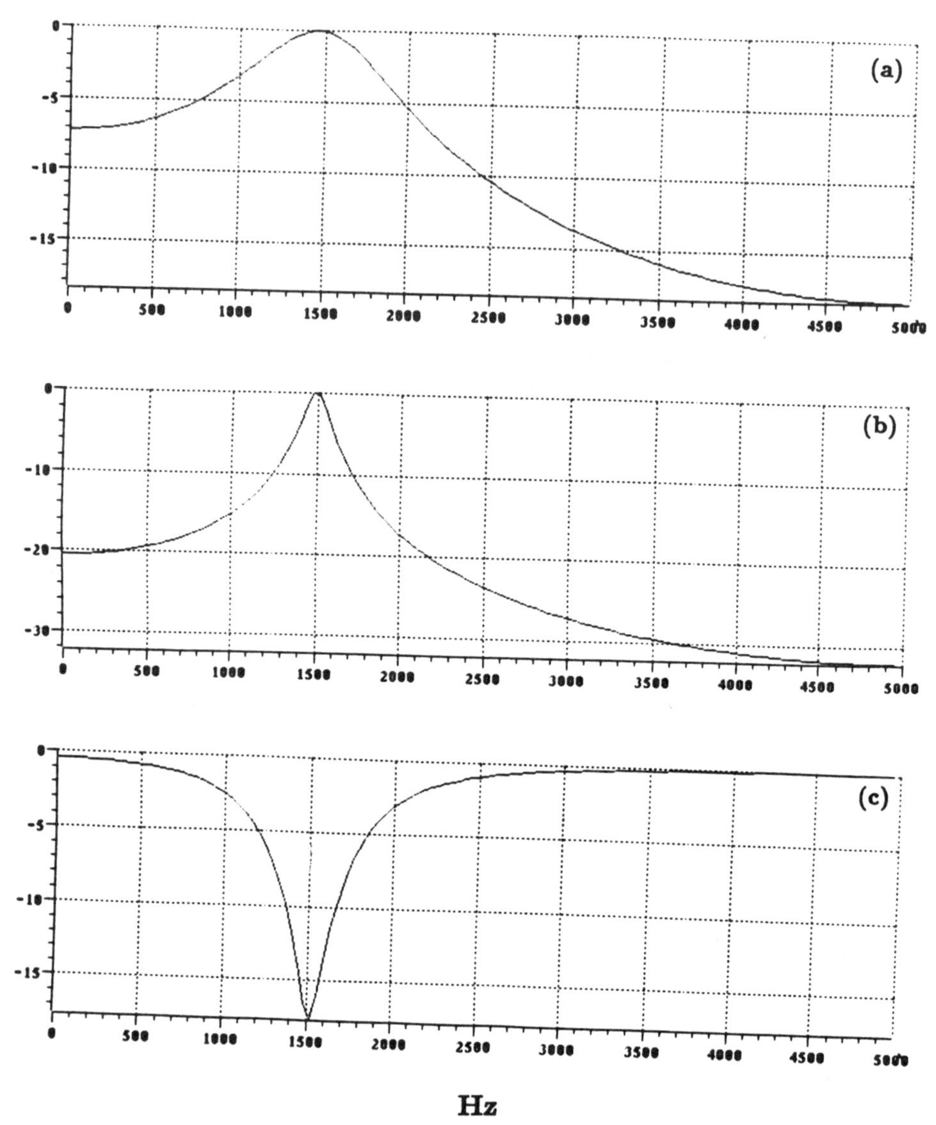

Figure 5.15. Transmission channels. (a) 750 Hz bandwidth pole at 1500 Hz (b) 150 Hz bandwidth pole at 1500 Hz. (c) Pole-zero pair at 1500 Hz of 1000 Hz and 150 Hz bandwidth, respectively.

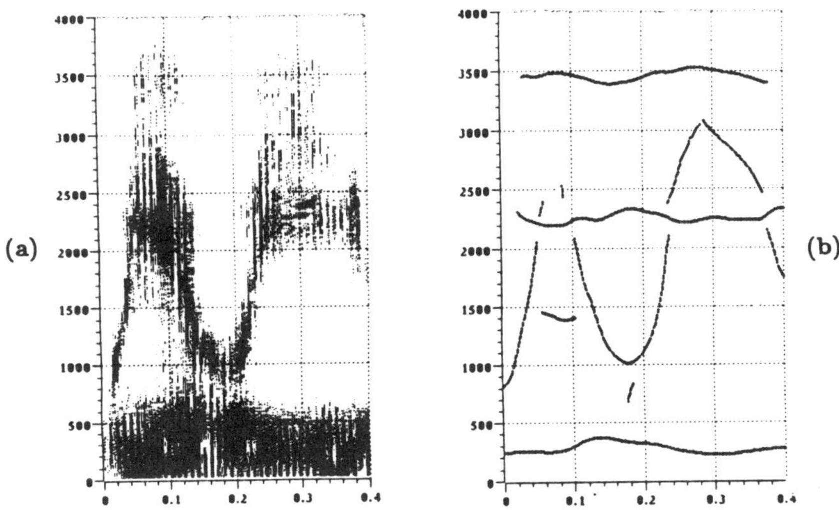

Figure 5.16. /wioi/ passed through tranmission channel in Figure 5.15a (broadband filter). (a) Wideband spectrogram. (b) Ridge analysis.

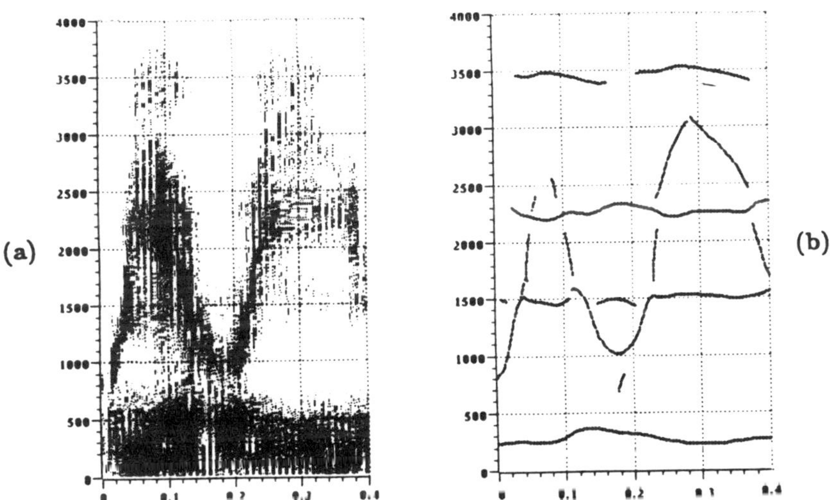

Figure 5.17. /wioi/ passed through tranmission channel in Figure 5.15b (narrowband filter). (a) Wideband spectrogram. (b) Ridge analysis.

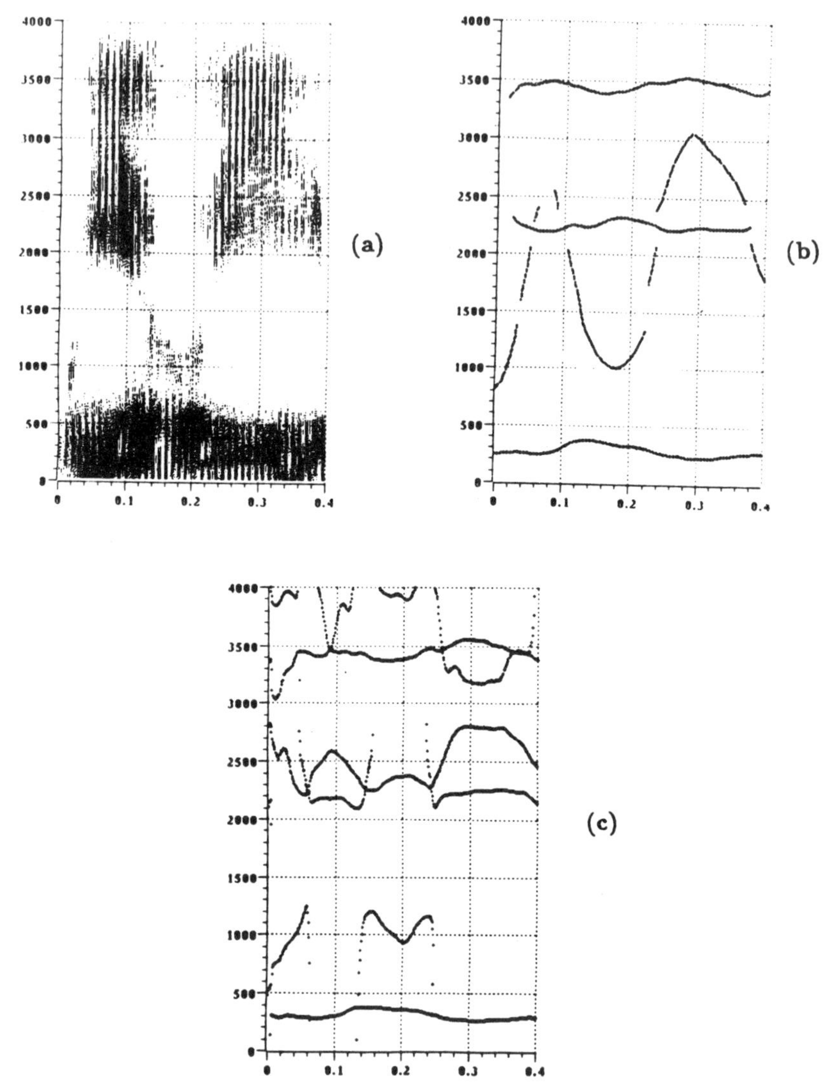

Figure 5.18. /wioi/ passed through tranmission channel in Figure 5.15c (notch filter). (a) Wideband spectrogram. (b) Ridge analysis. (c) LPC analysis.

References

Atal, B. & Hanauer, S. 1971. Speech analysis and synthesis by linear prediction of the speech wave. *J. Acoust. Soc. Am.* **50**, 637-655.

Bell, C., Fujisake, H., Heinz, J., Stevens, K., and House, A. 1961. Reduction of speech spectra by analysis-by-synthesis techniques. *J. Acoust. Soc. Am.* **33**, 1725-1736.

Beranek, L. 1954. *Acoustics.* New York: McGraw-Hill.

Bouachache, B., Escudié, B., Flandrin, P., & Gréa J. 1979. Sur une condition nécessaire et suffisante de positivité de la représentation cojointe en temps et fréquence des signeaux d'énergie finie. *C. R. Acad. Sciences.* **288. Serie A**, 307-309.

Bracewell, R. 1978. *The Fourier Transform and its Applications.* New York: McGraw-Hill.

Britt, R. & Starr, A. 1976. Synaptic events and discharge patterns of cochlear nucleus cells. II. Frequency-modulated tones. *J. Neurophysiol.* **39**. 179-194.

Chiba, T., & Kajimaya, M. 1941. *The Vowel, Its Nature and Structure*. Tokyo: Tokyo-Kaiseikan.

Claasen, T. & Mecklenbräuker, W. 1980a. The Wigner distribution: a tool for time-frequency signal analysis. Part I. *Philips J. of Res.* **35**, 217-250.

Claasen, T. & Mecklenbräuker, W. 1980b. The Wigner distribution: a tool for time-frequency signal analysis. Part II. *Philips J. of Res.* **35**, 267-300.

Claasen, T. & Mecklenbräuker, W. 1980c. The Wigner distribution: a tool for time-frequency signal analysis. Part III. *Philips J. Res.* **35**, 372-389.

Claasen, T. & Mecklenbräuker, W. 1984. On the time-frequency discrimination of energy distributions: can they look sharper than Heisenberg? *Proc. ICASSP '84.* **3**. 41B.4.1-4.

Cohen, L. 1966. Generalized phase-space distribution functions. *J. of Math. Phys.* **7**. 781-786.

De Bruijn, J. 1967. Uncertainty principles in fourier analysis. *Inequalities.* Shisha, O. (Ed.), New York: Academic Press. 57-71.

Do Carmo, M. 1976. *Differential Geometry of Curves and Surfaces.* Englewood Cliffs, NJ: Printice-Hall.

Dudgeon, D. 1984. Detection of narrowband signals with rapid changes in center frequency. *ICASSP DSP Workshop.* October 8-10. Chatham, MA.

Fant, G. 1960. *Acoustic Theory of Speech Production.* Hague: Mouton.

Fant, G. 1980. The relations between area functions and the acoustic signal. *Phonetica* **37.** 55-86.

Flanagan, J. 1956. Automatic extraction of formant frequencies from continuous speech. *J. Acoust. Soc. Am.* **28.** 110-118.

Flanagan. J. 1972. *Speech Analysis, Synthesis and Perception.* 2nd ed. New York: Springer-Verlag.

Flandrin, P. 1984. Some features of time-frequency representations of multicomponent signals. *Proc. ICASSP '84.* **3.** 41B.4.1-4.

Greenewalt, C. 1968. *Bird Song: Acoustics & Physiology.* Washington, DC: Smithsonian Institution Press.

Hille, E. 1984. *Functional Analysis and Semi-groups.* New York: Amer. Math. Soc.

Hlawatsch, F. 1984. Theory of bilinear time-frequency signal representations. *ICASSP DSP Workshop.* October 8-10. Chatham, MA.

Janssen, A. 1982. On the locus and spread of psuedo-density functions in the time-frequency plane. *Philips J. Res.* **37.** 79-110.

Jospa, P. 1982. Theory of evolving normal modes and the vocal tract. Part I: Basic relations and adiabatic invariance. *Acustica.* **52.** 86-94.

Jospa, P. 1984. Theory of evolving normal modes and the vocal tract. Part II: Evolving frequency and amplitude. *Acustica.* **57.** 133-138.

Kay, R. & Matthews, D. 1972. On the existence in human auditory pathways of channels selectively tuned to the modulation present in frequency-modulated tones. *J. Physiol.* **225.** 657-677

Kuhn, G. 1975. On the front cavity resonance and its possible role in speech perception. *J. Acoust. Soc. Am.* **58.** 428-433.

Ladefoged, P. 1975 *A Course in Phonetics*. New York: Harcourt, Brace, and Jovanovich.

Liberman, A., Cooper, F., Shankweiler, D., Studdert-Kennedy, M. 1967. Perception of the speech code. *American J. of Psychology*. **65**. 487-516.

Liberman, A., Delatter, P., Gerstman, L. 1954. The role of consonant-vowel transitions in the perception of the stop and nasal consonants. *Psychological Monographs*. **68**. 1-13.

Licklider, J. & Miller, G. 1951. The perception of speech. *Handbook of Experimental Psychology*. Stevens, S. (Ed.) New York: Wiley.

Liporace, L. 1975. Linear estimation of non-stationary signals. *J. Acoust. Soc. Am.* **58**. 1288-1295.

Lui, S. 1971. Time-varying spectra and linear transformation. *Bell Sys. Tech. J.* **50**. 2365-2374.

Loynes, M. 1968. On the concept of the spectrum for nonstationary processes. *J. Roy. Statist. Soc. (B)*. **30**. 1-20.

Markel, J. & Gray, A. 1976 *Linear Prediction of Speech*. New York: Springer-Verlag.

Marler, P. 1977. The structure of animal communication sounds. *Recognition of Complex Acoustic Signals*. Bullock, T. (Ed.) West Berlin: Dahlem Konferenzen.

Marr, D. 1982. *Vision*. San Francisco: Freeman.

Marr, D. & Hildreth, E. 1980. Theory of edge detection. *Proc. R. Soc. Lond. (B)*. **207**. 187-217.

Marr, D. & Nishihara, H. 1978. Representation and recognition of the spatial organization of three-dimensional shapes. *Proc. R. Soc. Lond. (B)*. **200**. 269-294.

Martin, W. & Flandrin, P. 1985. Wigner-Ville spectral analysis of nonstationary processes. *IEEE Trans. ASSP* **ASSP-33** 1461-1470.

Matthews, M., Miller, J. & David, E. 1961. Pitch synchronous analysis of voiced sounds. *J. Acoust. Soc. Am.* **45**, 458-465.

McCandless, S. 1974. An algorithm for automatic formant extraction using linear prediction spectra. *IEEE Trans. ASSP* **ASSP-22:2**. 135-174

Møller, A. 1978. Coding of time-varying sounds in the coclear nucleus. *Audiology*. **17**. 446-468.

Morse, P. & Ingard, K. 1968. *Theoretical Acoustics*. New York: McGraw-Hill.

Neuweiler, G. 1977. Recognition mechanisms in echolocation in bats. *Recognition of Complex Acoustic Signals*. Bullock, T. (Ed.) West Berlin: Dahlem Konferenzen.

Olive, J. 1971. Automatic formant tracking in a Newton-Raphson technique. *J. Acoust. Soc. Am.* **50**, 661-670.

Oppenheim, A. 1969. A speech analysis-synthesis system based on homomorphic filtering. *J. Acoust. Soc. Am.* **45**. 458-465.

Oppenheim, A. & Schafer, R. 1975. *Digital Signal Processing*, Englewood Cliffs, NJ: Prentice-Hall.

Page, C. 1952. Instantaneous power spectra. *J. Applied Phys.* **23**. 103-106.

Priestley, M. 1965. Evolutionary spectra and non-stationary processes. *J. Roy. Statist. Soc. (B).* **27.** 204-229.

Rabiner, L. & Schafer, R. 1978. *Digital Processing of Speech Signals.* Englewood Cliffs, NJ: Prentice-Hall.

Regan, D. & Tansley B. 1979. Selective adaptation to frequency-modulated tones: Evidence for an information-processing channel selectively sensitive to frequency changes. *J. Acoust. Soc. Am.* **65.** 1249-1257.

Riley, M. 1983. Schematizing spectrograms for speech recognition. *BTL Technical Memo. 11225-83-0924-07.*

Riley, M. 1984. Detecting time-varying spectral energy concentrations. *ICASSP DSP Workshop.* October 8-10. Chatham, MA.

Rudin, W. 1973. *Functional Analysis.* New York: McGraw-Hill.

Saleh, B. & Subotic, N. 1985. Time-variant filtering of signals in the mixed time-frequency domain. *IEEE Trans. ASSP.* **ASSP-33.** 1479-1485.

Schafer, R., & Rabiner, L. 1970. System for automatic formant analysis of voiced speech. *J. Acoust. Soc. Am.* **47.** 634-648.

Siebert, W. 1956. A radar detection philosophy. *IRE Trans. Inform. Theory.* **PGIT-6.** 204-221.

Stevens, K. & House, A. 1955. Development of a quantitative description of vowel articulation. *J. Acoust. Soc. Am.* **27.** 484-493.

Van Trees, H. 1968. *Detection, Estimation, and Modulation Theory, Pt. 1.* New York: Wiley.

Van Trees, H. 1971. *Detection, Estimation, and Modulation Theory, Pt. 3.* New York: Wiley.

Waltz, D. 1975. Understanding line drawings of scenes with shadows. *Psychology of Computer Vision.* Winston, P. (Ed.), New York: McGraw-Hill.

Zadeh, L. 1950. Frequency analysis of variable networks. *Proc. IRE.* **38.** 291.

Index